Geoengineering Discourse Confronting Climate Change

Geoengineering Discourse Confronting Climate Change

The Move from Margins to Mainstream in Science, News Media, and Politics

Brynna Jacobson

LEXINGTON BOOKS
Lanham • Boulder • New York • London

Part of chapter 2 previously published. Brynna Jacobson, 2018, "Constructing Legitimacy in Geoengineering Discourse: The Politics of Representation in Science Policy Literature." *Science as Culture* 27(3): 322–348.
Part of chapter 3 previously published. Brynna Jacobson, 2018, "Constructing Legitimacy in Geoengineering Discourse: The Politics of Representation in Science Policy Literature." *Science as Culture* 27(3): 322–348.

Published by Lexington Books
An imprint of The Rowman & Littlefield Publishing Group, Inc.
4501 Forbes Boulevard, Suite 200, Lanham, Maryland 20706
www.rowman.com

6 Tinworth Street, London SE11 5AL, United Kingdom

British Library Cataloguing in Publication Information Available

Library of Congress Cataloging-in-Publication Data Available

ISBN 978-1-7936-3528-0 (cloth : alk. paper)
ISBN 978-1-7936-3530-3 (paper : alk. paper)
ISBN 978-1-7936-3529-7 (electronic)

♾™ The paper used in this publication meets the minimum requirements of American National Standard for Information Sciences—Permanence of Paper for Printed Library Materials, ANSI/NISO Z39.48-1992.

Dedicated to my parents, Addie and Paul.

Contents

SECTION IV: CONGRESSIONAL HEARINGS   173

SECTION V: TECHNOLOGY AND REFLEXIVITY   219

Acknowledgments

This has been the work of many years and has benefited from the help and support of many people.

I would like to thank all the members of my doctoral committee at the University of California, San Diego, who saw me through the dissertation stage of this research: Kelly Gates, Isaac Martin, Bud Mehan, Kwai Ng, and especially my committee chair, Charles Thorpe, who helped me choose the topic of geoengineering in the first place and supported and encouraged me throughout the research journey.

This book is dedicated to my parents, Addie and Paul Jacobson, who have always been there for me. In respect to this project in particular, I don't know how I would have done it without my mom who read countless chapter drafts and provided valuable input.

My husband, Marek Adamo, has supported me throughout every stage of this project and been the best partner I could ask for. I would like to thank him for being a wonderful husband and father and, of course, for graciously accepting his perpetual role as tech support.

I am thankful to our children, Corinne and Daniel, for providing me with motivation, purpose, and perspective. While they did not particularly make writing any easier, I love them all the more for it!

Thank you to everyone who helped with childcare at various stages while I researched and wrote, especially my children's grandparents, Addie and Paul Jacobson and Marilyn and Alan Adamo; my sister-in-law Lauren Adamo; and my friend Leda Bashi.

To my whole family and dear friends, thank you for your love and support in this venture over many years. In particular, I want to acknowledge my siblings, Willow Jacobson and Quinn Jacobson, who have both inspired me my whole life and whom I still go to for advice. I would also like to honor the memory of my grandparents, all four of whom valued education and research and passed those values to the generations that followed.

I want to thank Courtney Morales at Lexington Books who has been enthusiastic about this manuscript from the beginning and patient as world events, including a global pandemic, changed schedules. The final version of this work also benefited from the insightful comments made through anonymous peer review for which I am grateful.

Section II includes content that was originally published as an article in *Science as Culture*, and I am grateful to the editors and anonymous reviewers who provided input on that original article. Substantial updates have been made in this manuscript, however, to reflect new developments in the field.

Thanks to Nicole Hill Arik for expertly and awesomely realizing my vision for the cover art: the earth constructed of the discourse of geoengineering.

I started this project as a graduate student in the Sociology Department at the University of California, San Diego, and have been teaching in the Sociology Department at the University of San Francisco in the years since. I am grateful for the wonderful people in both departments and for the sense of community they have given me.

Acronyms

AM	Albedo Modification
BECCS	Bioenergy with carbon capture and sequestration
CCT	Cirrus Cloud Thinning
CDR	Carbon Dioxide Removal
DAC	Direct air capture
GHG	Greenhouse gas
IPCC	Intergovernmental Panel on Climate Change
MCB	Marine cloud brightening
NASEM	National Academies of Sciences, Engineering, and Medicine
NETs	Negative Emissions Technologies
NRC	National Research Council
SAAM	Stratospheric Aerosol Albedo Modification
SAI	Stratospheric aerosol injection
SCoPEx	Stratospheric Controlled Perturbation Experiment
SG	Solar geoengineering
SPICE	Stratospheric Particle Injection for Climate Engineering
SRM	Solar Radiation Management
STS	Science and Technology Studies
UK	United Kingdom
U.S.	United States

Section I

CLIMATE, GEOENGINEERING, RISK, AND MODERNITY

Introduction

Climate Crisis, Global Risk, and Geoengineering

INTRODUCTION

Climate change with its associated global risks is a quintessential, and existential, dilemma of "high modernity" (e.g., Beck 2009; Giddens 1990; Thorpe and Jacobson 2013). As such, how global society addresses climate change and the processes through which this is determined are crucial. To date, with few exceptions, climate politics have been exemplary of the "politics of unsustainability," characterized by "general acceptance that the achievement of sustainability requires radical change in the most basic principles of late-modern societies" and a simultaneous presumption that such radical change is politically impossible (Blühdorn and Welsh 2007, 198; Blühdorn 2011; Blühdorn 2007). In regard to climate change and global politics, "sustainability" has been all but absent from the discourse, with the narrower notion of "mitigation" declared the objective, as in mitigating the range of possible temperature change through reducing emissions and mitigating the human toll of climate change through adaptation efforts. Since 2006, the already contentious politics around climate change have been further complicated by the increasing attention given to geoengineering among the "portfolio of options" to address climate change (National Research Council 2015b; National Research Council 2015a).

Distinct from mitigation efforts that would adjust human and economic behavior in light of anthropogenic climate change, geoengineering includes a range of "techno-fix" proposals intended to modify the functioning of the climate itself. Recurrent climate summit negotiation failures have created the political conditions for climate engineering to emerge within scientific, political, and media discourse as a potential "Plan B." However, with "Plan A" (i.e., emissions reductions) at odds with entrenched political-economic interests so

3

that it has not gained full commitment from governments beholden to powerful economic interests (Foster 2011; Foster, York and Clark 2009), a gap has been left through which Plan B has begun encroaching upon Plan A.

As climate scientist James Hansen unequivocally contends, climate change puts Earth "in imminent peril" (Hansen 2009, ix). To date, however, international efforts to address climate change and mitigate its effects have failed by most accounts. There is an overwhelming indication that the primary way to address climate change must be through limiting carbon emissions, primarily by reducing use of fossil fuels (e.g., National Research Council 2015b, 17; Royal Society 2009, ix; IPCC 2014, 17–18; Melillo, Richmond and Yohe 2014, 13, 649; Wuebbles et al. 2017, 31–32). As science journalist Michael Specter succinctly articulates the dilemma in the *New Yorker*:

> The best solution, nearly all scientists agree, would be the simplest: stop burning fossil fuels, which would reduce the amount of carbon we dump into the atmosphere. That fact has been emphasized in virtually every study that addresses the potential effect of climate change on earth—and there have been many—but none have had a discernible impact on human behavior or government policy. (Specter 2012, 100)

The challenge of climate change, and employing the obvious solution of emissions reductions, has been particularly intractable due to the extent to which carbon emissions are a core externality of the global energy economy. Adequately reducing emissions would require extensive restructuring of the energy economy and all carbon-intensive industries, a course which has been thoroughly challenged by powerful political-economic interests (Schnaiberg 1980; Foster 2002; Gould, Pellow and Schnaiberg 2008; McCright and Dunlap 2010; Urry 2011; Harris 2013; Klein 2014; Gunderson, Stuart and Petersen 2018; Saspinski, Buck and Malm 2021).

Describing the long-experienced reluctance of nations to make meaningful efforts on containing emissions as a kind of tragedy of the commons, political scientist Paul Harris contends: "The problem lies in the convenient but pernicious reality that everyone is free to use the global atmosphere as a dumping ground [. . .] everyone is free to pollute the atmosphere, and we have done so with abandon for hundreds of years" (2013, 4). For most of industrial history carbon pollution has been entirely an externality with no real costs incurred by those creating the pollution. This basic tenet of industrial development has shaped the formation of our energy economy and all the infrastructure, goods, and lifestyles surrounding it.

A further challenge to emissions reductions has been the competitive posturing of nation-states, which have been the primary unit involved in negotiations (Harris 2013). Many nations, especially the largest polluters, have been loath to incur economic competitive disadvantages or the costs that would be associated with reductions, causing a stalemate in reduction efforts for several decades (e.g., Harris 2013; Giddens 2009, 14; Victor 2011, xxx, 62, 263; Foster 2002, 13–16). Political scientists Frank Biermann and Klaus Dingworth explain that "global environmental change increases the mutual dependence of nation states, thereby further undermining the idea of sovereignty as enshrined in the traditional Westphalian system" (Biermann and Dingworth 2004, 2). As a result of mutual dependence—and impact—on the environment, "the modern complex ecological interdependence binds all nations, which creates a new dependence of individual nation states—even the largest, most powerful ones—on the community of all other nations" (Biermann and Dingworth 2004, 6). The dominant framework of sovereignty and realpolitik driven by national interests and competition, however, obstructs these new "interdependencies" from finding cooperative articulation. In this way, climate change expresses a contradiction between the global character of production, including its environmental impacts, and the division of the world into nation-states as the units with primary governing authority.

For almost two decades following the 1997 Kyoto Protocol, international summits aimed at emissions reductions failed to secure meaningful agreements or even a significant semblance of progress (Victor 2011; Blühdorn 2011; Harris 2013). Until the 2015 Paris Accord, the post-Kyoto climate summits have been considered resounding failures by almost any measure. The Paris Summit may be the exception that proves the rule. The Paris conference in late 2015, widely heralded as the most successful negotiation to date, resulted in a nonbinding (until further ratified) agreement to curb warming through mitigation. However, as reported in the *New York Times*:

> The new deal will not, on its own, solve global warming. At best, scientists who have analyzed it say, it will cut global greenhouse gas emissions by about half enough as is necessary to stave off an increase in atmospheric temperatures of 2 degrees Celsius or 3.6 degrees Fahrenheit. (Davenport 2015)

Consequently, it has become clear that the ambitions of the Paris Agreement rely upon a presumption of geoengineering, at least in the form of Carbon Dioxide Removal (CDR) or so-called "negative emissions," as a necessary component (Craik and Burns 2019; Horton, Keith and Honegger 2016; IPCC 2018). Andreas Malm critiques the Paris Agreement targets as "likely beyond reach," with the planet "already doomed 'to a mean warming over land greater than 1.5°C'" (Malm 2018, 9; citing Huntingford and Mercado 2016). Indeed, the Intergovernmental Panel on Climate Change (IPCC) has subsequently indicated an expected overshoot of the 1.5 degrees Celsius goal, with even the most optimistic "very low emissions scenario" likely to surpass

1.5 degrees Celsius before declining, and all other scenarios expected to fail to meet the goal by end of century (IPCC 2021, 18). The diplomatic success of the Paris Accord, and the prospects of an emissions scenario playing out that would hold warming within 2 degrees Celsius, still face the challenge of nations actually delivering on their pledged goals in face of the recurrent challenges to emissions reductions.

Looking at the critical example of the United States, the world's largest cumulative and second largest annual carbon-emitting nation, major challenges to mitigation have been political and socio-cultural. Sociologists Aaron McCright and Riley Dunlap argue that the American conservative movement has effectively used the "second dimension of power" to confine the scope of climate policy decision-making within the federal government "to only those issues that do not seriously challenge their subjective interests" (McCright and Dunlap 2010, 106; see also Lukes 2005). This has included challenging impact science and scientists, invoking favorable political procedures, and affecting public opinion through invoking media bias (McCright and Dunlap 2010). This political analysis fits with other trends of cultural staging of climate risks. Since perception affects how the public responds to risks, the ways in which risks are staged and defined have real effects upon the potential unfolding of that risk (Beck 2009, 16, 20).

Historians Naomi Oreskes and Erik Conway (2010) find that shaping public perception of risks has become a profitable business strategy. In *Merchants of Doubt*, they examine the manipulation of science, message, and media used to distort public perceptions of risks. The tobacco industry discovered and capitalized on the strategy of invoking uncertainty to sow doubt by using "*normal* scientific uncertainty to undermine the status of actual scientific knowledge" (Oreskes and Conway 2010, 34, emphasis in original). While tobacco developed this art, it persists today with the most striking example being the disproportionate impact of climate change denial. Tracing the historical circumstances that contributed to the downplaying of global warming in the decades following its discovery, Oreskes and Conway underscore the manipulation of and by the media that has contributed to public confusion and doubt (2010, 170–215). By providing "fair and balanced" presentation of climate change with equal attention to both the majority scientific view and the minority dissenting view, as if it was an issue of scientific controversy, the media has distorted the impression of climate change in favor of skeptics and deniers (Oreskes and Conway 2010, 214; cf., Perrow 2011, xxxviii).

Frederick Buell similarly refers to the "enormously successful anti-environmental disinformation industry" that sprang up in opposition to the environmental trends that emerged in the 1970s (2003, 3). This new industry helped midwife a new phase in the history of U.S. environmental politics, one in which an abundance of environmental concern was nearly blocked by an equal abundance of antienvironmental contestation. [. . .] Despite scientific evidence and even, in a number of cases, virtual scientific consensus to the contrary, issue after issue was contested. (Buell 2003, 4)

The contested issues included the ozone hole, food and population crises, the effects of chemical pesticides like DDT, and global warming. Buell identifies the pseudoscience performed by conservative, ideologically driven think tanks and others as the "counterscience movement." This movement has been "devoted to countering the findings of environmental science with the creation of a body of antienvironmental science," which is not "science" in the conventional sense but models itself in the guise of academic-style articles filled with statistics and documentation, which are often unreliable, and "references to fellow counterscience writers, most of whom were not scientists but anti-environmental journalists, economists, and ideologues" (Buell 2003, 5). Buell places these antienvironmental efforts within the context of the new conservative movement that arose in the 1980s.

This movement continues today, now institutionalized as a mainstream political influence with successful entrenchment within state politics and the U.S. Federal Government, as demonstrably manifested within the executive branch during the presidencies of George W. Bush and Donald Trump as well as within Congress. Despite the near unanimity among the scientific community (Oreskes 2004), politically climate change has been the casualty of "heightened partisan polarization" (Dunlap, McCright and Yarosh 2016; cf. Mehling and Vihma 2017). Climate change has been systematically thwarted from achieving unhampered attention, let alone meaningful mitigation policies. By maintaining a semblance of doubt, deniers reframe the issue and divert politics away from solutions. Moreover, politicians are able to express doubt, premised on supposedly ongoing debates regarding the existence of climate change, and thereby challenge mitigation at various levels of government.

Resulting from the successful entrenchment of the conservative antienvironmental movement within government, one obstacle to fulfilling the U.S. commitments to the Paris Agreement has been internal polarization and contention within the U.S. polity, with various disputes arising between state governments, Congress, the Supreme Court, and the Executive branch. For example, in response to mitigation policy enacted by the Obama administration, 29 states jointly filed lawsuit in 2015 against the most important components of the federal efforts to reduce greenhouse gases (GHGs) in the case *State of West Virginia, State of Texas, et al. v. United States Environmental Protection Agency*. In this case, the Supreme Court unprecedentedly issued a stay on the new federal regulations for coal-fired power plants "before review by a federal appeals court" (Liptak and Davenport 2016). Such challenges aside, some independent assessments conclude "even if the [Obama] administration [had] executed all its existing and planned policies with maximum effect, and the most optimistic forecasts for technological development and forest sink capacity were borne out, the United States would still not hit the target" (Porter 2016).

Turning the federal-state dynamic on its head and emphasizing the deep political polarization—and hence political precariousness—that has developed in the United States around climate change, the executive transition from the Obama administration to the Trump administration led to an abrupt reversal on climate policy, exacerbating long-standing challenges to mitigation and stifling newfound international momentum. The Trump administration pursued a general crusade against environmentalism, marking a pointed reversal of the executive branch from fostering climate mitigation policy to proactively opposing it. The seriousness of this shift was signaled from the outset in terms of both policy and political appointees that demonstrably signaled a drastic shift toward unveiled antienvironmentalism in rhetoric, representation, and policy. Climate change denier Scott Pruitt was initially placed in charge of the Environmental Protection Agency (EPA), later succeeded by Andrew Wheeler, a former coal industry lobbyist. Likewise, for the role of secretary of state, the official face of international relations and diplomacy, Trump exclusively appointed men who embody the oil and gas industry, first with Rex Tillerson, former Exxon Mobil executive, followed by Mike Pompeo, whose career has been defined by association with Koch Industries (Janetsky and Kelly 2018).

Within his first month in office, Trump declared his intention to spur coal, oil, and gas extraction and commerce in the United States, cut investment in climate change research, reduce environmental regulation, and defund the EPA. Dismantling environmental protections and policies was a centerpiece of the administration, with over 100 environmental rules, regulations, and commitments dismantled or rolled back (Popovich, Albeck-Ripka, and Pierre-Louis 2020). As both a symbolic and material embodiment of this antienvironmentalism, in August 2020, plans were solidified to "open up part of the Arctic National Wildlife Refuge in Alaska to oil and gas development, a move that overturns six decades of protections for the largest remaining stretch of wilderness in the United States" (Plumer and Fountain 2020). While the administration actively sought to widely overturn environmental protections and promote oil and gas interests generally, it was particularly keen to undermine climate mitigation and especially the advancements made by the Obama administration, both its domestic efforts such as the Clean Power Plan, and international achievements, including the benchmark 2015

Paris Agreement on climate change, from which Trump served an intent to withdraw notice in November 2019.

The U.S. Federal Government's reversal on climate policy shaped a new global political context. At least rhetorically, the response to the Trump administration's actions on climate was an upswell in international commitment to climate mitigation efforts. For example, following the G20 Summit in July 2017, the 19 members other than the United States reaffirmed commitment to climate mitigation and specifically to the Paris Accord (see European Commission 2017). Domestically in the United States, increased mitigation efforts have percolated up from local and state levels. The United States Climate Alliance, founded by the states of California, New York, and Washington in June 2017, has grown to include 25 states and territories as of July 2020 (United States Climate Alliance 2020). Participating states are "committed to achieving the U.S. [Paris Accord] goal of reducing emissions 26–28 percent from 2005 levels and meeting or exceeding the targets of the [Obama era] federal Clean Power Plan" irrespective of federal policy (Inslee 2017). In the face of uncertain U.S. federal policy, these local, state, and international actors proceeded down the historically challenging path of instituting carbon mitigation commitments.

With the transfer of power again in 2021, the United States marked yet another dramatic climate policy reversal at the federal level as the Biden administration set out with the intent to push the climate agenda further than past administrations have achieved. During the transition period, Biden made grand overtures regarding his intent for climate to be a "cornerstone" of his administration, including the nomination and naming of a diverse and vibrant "climate team," which the president-elect said would be "ready on day one" (Friedman 2020). On his first day in office, President Biden recommitted the United States to the Paris Agreement and on January 27 took additional executive actions related to prioritizing climate change (White House 2021b), signaling the administration's commitment to making climate a core priority throughout departments and agencies. Efforts at climate legislation, however, have continued to face significant opposition in Congress, and, as of the time of writing, the signaled seriousness regarding the monumental task of addressing climate change has not been matched with correspondingly serious policy.

To date, a central challenge in facing climate change has been the difficult reality that drastically reducing carbon emissions in response to climate change requires major adjustments to the political-economic system as a whole. Political scientist David Victor argues:

Tinkering at the margins of the energy system won't make much of a difference. Deep cuts in CO_2 will probably require a massive re-engineering of modern energy systems. [. . .] And because this transformation will require new technologies and business models that do not yet exist the political interest groups that can keep the process on track do not yet exist. (Victor 2011, 4)

As such, political will to address the problem in a meaningful way has been conspicuously lacking. For his part, Victor argues for a rethinking of diplomatic approaches to be more "realistic" and also for research into geoengineering (Victor 2011, 242, 21). Geoengineering, which adds a new layer to the "politics of unsustainability" (Blühdorn 2007; 2011), will be discussed presently.

GEOENGINEERING

In the face of increasingly visible effects of climate change and continued challenges to mitigation efforts, the notion of geoengineering (also called "climate engineering" or "climate intervention") has shifted toward mainstream attention in recent years. As defined by the IPCC: "Geoengineering refers to a broad set of methods and technologies operating on a large scale that aim to deliberately alter the climate system in order to alleviate the impacts of climate change" (IPCC 2014, 89). Geoengineering reflects a culture of technological exuberance characterized by optimism in the ability of technology and innovation to overcome environmental challenges, including those problems created themselves by technology and innovation (cf. Huesemann and Huesemann 2011). Distinct from mitigation efforts to adjust human and economic behavior, geoengineering "techno-fix" proposals aim to modify functioning of the climate itself, involving scales of technology with profound planetary effects and extensive risks.

The umbrella term "geoengineering" (also known as "climate engineering" or "climate intervention") encapsulates various proposals to modify climate through intentional manipulation. There are two main categories of geoengineering aimed at addressing climate change: (1) albedo modification, also known as "solar geoengineering" (SG), "albedo enhancement," "solar climate engineering," or "solar radiation management" (SRM), and (2) carbon dioxide removal (CDR), also known as Negative Emissions Technologies (NETs). There have been other proposals that could be considered geoengineering,[1] but for the most part, albedo modification and CDR are the primary categories of geoengineering proposals that have been shifting toward serious consideration. Albedo modification would involve "intentional efforts to increase the amount of sunlight that is scattered or reflected back to space, thereby reducing the amount of sunlight absorbed by the Earth" in order "to produce a cooling designed to compensate for some of the effects of warming associated with greenhouse gas increases" (National Research Council

2015b, 25, 28). Various methods have been proposed to achieve such albedo modification (i.e., increase the degree to which sunlight is reflected) such as releasing sulfate particles in the stratosphere (known as stratospheric aerosol injection or SAI) or changing the properties or distribution of clouds (for example, through marine cloud brightening, MCB, or cirrus cloud thinning, CCT). In contrast, CDR proposals seek to capture atmospheric carbon on a massive scale and redirect it into various repositories, including soil, vegetation, oceans, or underground storage. The array of CDR proposals is broad and diverse, including direct air capture (DAC) through machines and chemical processes to extract carbon from the ambient air, enhanced weathering that would increase the carbon absorption of minerals, and bioenergy with carbon capture and sequestration (BECCS) in which biomass is processed from energy production with the resulting CO_2 to be captured as a point source emission.

The modern idea of science-based weather modification has been pursued at least since the 1960s (National Research Council et al. 2003, 1; Goodell 2010, 70; Fleming 2010), and the concept of geoengineering specifically in regard to climate change has a history dating back to 1965 when anthropogenic climate change first emerged as a policy issue. The 1965 Report of the Environmental Pollution Panel of the President's Science Advisory Committee presented a compelling overview of the greenhouse effect and projected climate change in coming decades, with the sole policy recommendation to President Johnson that the United States should be pursuing albedo modification as a possible response to climate change (Environmental Pollution Panel 1965, 126–127). At the time, the terminology had not yet been created, so there is no reference to "geoengineering" but rather conceptual descriptions of albedo modification:

> The climatic changes that may be produced by the increased CO_2 content could be deleterious from the point of view of human beings. The possibilities of *deliberately bringing about countervailing climatic changes* therefore need to be thoroughly explored. *A change in the radiation balance in the opposite direction* to that which might result from the increase in CO_2 could be produced by *raising the albedo, or reflectivity, of the earth.* (Environmental Pollution Panel 1965, 127, emphasis added)

The recommendation of albedo modification in this 1965 report as the solution to pursue for addressing climate change reflects the instrumental relationship with nature that characterized U.S. environmental politics prior to the emergence of the environmental movement that gained a clear presence about five years later (cf. Kirsch and Mitchell 1998; Lindseth 2013). The instrumental treatment of nature, and the idea of mastery over it, reflected in the 1965 report were in step with concurrent weather modification efforts in the United States, premised upon a human intention to "control" nature, a cultural conception with a long history (Fleming 2010, 3–10).

As with other technological ventures in the period since World War II, the U.S. military has been particularly involved in the pursuit of weather modification for strategic purposes (Fleming 2010, 165–188). In the same era that the 1965 Science Advisory report was recommending climate control through albedo enhancement to counteract global warming, the U.S. military was commencing use of weather modification for purposes of war. Cloud seeding to generate rain was used "between 1966 and 1972 in the jungles over North and South Vietnam, Laos and Cambodia" (Fleming 2010, 179). Despite public backlash regarding the use of weather control efforts in the Vietnam War (as exposed post hoc), at least 30 years later weather modification remained a component of military pursuit as demonstrated by a 1996 report to the Air Force, which claimed that by 2015 "US aerospace forces can 'own the weather' by capitalizing on emerging technologies and focusing development of those technologies to war-fighting applications" (House et al. 1996, vi).

While the initial Science Advisory report on climate change emphasized albedo modification, in subsequent years, at least since the 1992 United Nations Framework Convention on Climate Change, the emphasis of public political discourse on addressing climate change has focused on mitigation through emissions reductions and adaptation efforts. In recent years, however, there has been renewed interest in geoengineering strategies to address climate change through "deliberate, large-scale intervention in the climate system designed to counter global warming or offset some of its effects" (Hamilton 2013a, 1). Mainstreaming has occurred within scientific, political, and public spheres. The mounting evidence and concern regarding climate change, the limited scope of mitigation efforts to date, and social fascination with novelty and technological development have pushed the concept of geoengineering onto the radar of the broader public. Advocates for pursuing, or at least considering, geoengineering options include a subset of engineers, climate scientists, economic and political interests, and even environmentalists.[2]

Terminology has shifted over time in regard to geoengineering and its subtypes. The term "geoengineering" is attributed to a 1977 article by physicist Cesare Marchetti (Keith 2001, 497; Marchetti 1977). The term "solar radiation management" was coined by climate scientist Ken Caldeira in 2006 as an alternative to speaking of "geoengineering" (Hamilton 2013a, 76). Notably, in the benchmark 2015 National Research Council (NRC) of the National Academy of Sciences two-volume report on geoengineering, the authoring committee consciously changed the discourse through intentional terminology and word choice. They reject the terms "geoengineering" and "climate engineering," explaining that not only do these terms lack specificity but "the term 'engineering' implies a more precisely tailored and controllable process than might be the case for these climate interventions" (National Research Council 2015b, 1).

The 2015 NRC report also rejects the terms "solar radiation management" (SRM) and "albedo enhancement" in favor of the more neutral term "albedo modification." This choice of terminology is not explained except to say

the Committee chose to avoid the commonly used term of "solar radiation management" in favor of the more physically descriptive term "albedo modification" to describe a subset of such techniques that seek to enhance the reflectivity of the planet to cool the global temperature. (National Research Council 2015b, x)

Similar to avoiding the term "engineering," steering away from "SRM" signifies a move toward establishing more neutral terminology in regard to these technologies. Unlike "solar radiation management," the phrase "albedo modification" is descriptive without casting negative attribution to "solar radiation" (which is obviously essential to life on Earth) or implying feasibility of precisely controlling it as suggested by the term "management."[3] The follow-up report published in 2021 by the National Academies of Sciences, Engineering, and Medicine (NASEM) once again changes course on the terminology, adopting "solar geoengineering" (SG) in place of "albedo modification." With this change, the 2021 Committee continues the precedent of acknowledging the importance of conscientious reflection and thought regarding what terminology is used, stating:

These terminology issues are worthy of ongoing consideration as they represent more than a semantic debate; in fact, terminology can affect public perceptions and opinions of the various response strategies proposed and can help frame the discourse moving forward. (NASEM 2021, 21)

In recognition that various organizations, reports, and authors use different labels to describe the concepts, this manuscript will necessarily employ the various terms that have been used for these concepts interchangeably, with effort to reflect the original terminology within the sources being discussed whenever possible.

Albedo modification, the idea of proactively increasing the percentage of solar radiation reflected away from Earth, is the more controversial proposed form of geoengineering. The most influential science policy reports, including those by the Royal Society and NRC, recommend that research be pursued but that "Albedo modification at scales sufficient to alter climate should not be deployed at this time" (National Research Council 2015b, 7). Albedo modification is inherently high-risk. It would constitute a global experiment with large-scale risks of changing ecosystems, hydrological systems, regional climates, and other essential Earth systems (cf. Macnaghten and Szerszynski 2013; Stilgoe 2016; Owen 2014; Krohn and Weingart 1987; Levidow and Carr 2007). As broadly articulated by the NRC Committee, "Introducing albedo modification at scales capable of substantial reductions in climate impacts of future higher CO_2 concentrations would be introducing a novel situation into the Earth system, with consequences that are poorly constrained at present" (National Research Council 2015b, 7).

CDR, on the other hand, has been treated as less controversial and has been increasingly recommended for implementation by esteemed scientific organizations. In its seminal 2009 report, the Royal Society concluded that "Carbon Dioxide Removal methods that have been demonstrated to be safe, effective, sustainable and affordable should be deployed alongside conventional mitigation methods as soon as they can be made available" (Royal Society 2009, xi). In 2015, the second recommendation of the NRC report, after that of continuing mitigation efforts, is for "research and development investment to improve methods of carbon dioxide removal and disposal at scales that matter" (National Research Council 2015a, 91). In the Forward to the 2018 "Greenhouse Gas Removal" report, jointly written by the Royal Society and the Royal Academy of Engineering, the presidents of these prestigious scientific societies write: "It is increasingly clearer that reducing emissions is not enough" to meet the Paris Accord goal of containing warming to 2 degrees Celsius, but that "we must also actively remove greenhouse gases from the atmosphere" and that "action must begin now" (Royal Society and Royal Academy of Engineering 2018, 7).

As for the IPCC, the 2014 IPCC report markedly included CDR as an assumed element of carbon mitigation going forward. The authors note that "a large proportion of the new scenarios [in the report] include Carbon Dioxide Removal (CDR) technologies" (IPCC 2014, 21). The 2018 IPCC Special Report further embraces CDR, stating: "All pathways that limit global warming to 1.5°C with limited or no overshoot project the use of carbon dioxide removal (CDR) on the order of 100–1000 $GtCO_2$ over the 21st century" (IPCC 2018, 17). By the time of the 2021 IPCC Report, CDR and the related concept of "negative emissions" have become central presumptions of future climate policy (IPCC 2021).[4] Moreover, the report indicates that the Working Group III contribution, expected for completion in 2022, will further expound upon the importance of CDR.

While CDR is more widely accepted than albedo modification and generally expected to become part of the mitigation program along with emissions reductions efforts, it has not been immune to controversy and criticism. Historically, some proposed CDR methods have received criticism from environmentalists concerned about distraction from emissions reductions efforts as well as ecological harms that would be caused by the CDR techniques. For example, in the 1990s and 2000s, a CDR proposal that received significant attention was Ocean Iron Fertilization (OIF), which was based on the idea that adding minerals to the ocean could create algal blooms that would later sink to the bottom of the sea, bringing sequestered carbon with it. Scientific experiments began in the 1990s in numerous locations around the world, but backlash from environmental organizations and governments followed when attempts to scale and profit from OIF were pursued, in particular by Russ George and his Planktos Corporation, which attempted to pursue large-scale experiments at various sites including near the ecologically sensitive Galápagos Islands (Möller 2021, 24; cf. Schneider and Fuhr 2021, 63–64; cf. Goodell 2010, 143–158). Following this controversy, an amendment of the international agreement regarding marine pollution known as the London Protocol "prohibited ocean fertilization for nonscientific purposes" with decisions of the U.N. Convention on Biological Diversity following suit (Möller 2021, 24–25). Land-based CDR proposals, such as bioenergy with carbon capture and sequestration (BECCS), have also been critiqued as problematic due to the extensive land use they would require, which would necessarily divert significant land away from agriculture and natural ecosystems, resulting in a loss of land dedicated to food cultivation and biodiversity (e.g., see Malm 2021, 153).

Recently, the elevation and entrenchment of CDR in climate models and political and scientific discourse has been subject to critique by social scientists concerned with its increasingly assumed role in climate policy. In their critical social analysis of CDR, Diana Stuart, Ryan Gunderson, and Brian Petersen find "limited possibilities for carbon geoengineering to contribute toward mending the carbon rift especially in our current social metabolic order" (Stuart, Gunderson and Petersen 2020, 1243). Wim Carton (2021) questions the indication that "negative emissions" are feasible, let alone necessary, to achieve climate goals. Carton argues that NETs, which are now deeply embedded in the climate models the IPCC uses to inform climate policy making, were not an inevitable outcome or conclusion, but rather that "the inclusion of negative emissions in models and from there into the IPCC was profoundly ideological" and that "negative emissions are already being invoked to justify business as usual" (Carton 2021, 36). However, many critics of CDR, including those critiquing the underpinning ideologies, ecological risks, unrealistic promise of scope and feasibility, and concerns of moral hazard in regard to emission reduction efforts, acknowledge that CDR has a role to play in climate policy. As Andreas Malm admits, "even with the most formidable plans for radical emission cuts" CDR will be necessary "*because too much has already been emitted*" (Malm 2021, 155, emphases in original).

However, accepting that negative emission technologies must have a role to play in climate mitigation does not negate legitimate critiques of CDR proposals and the socio-technical imaginaries of how they play out, nor the overarching issue that CDR remains emblematic of a persisting paradox of approach in addressing climate change. The axiom, attributed to Benjamin Franklin, that "an ounce of prevention is worth a pound of cure" is apt to understanding this paradox of pursuing carbon dioxide removal technologies while continuing to burn fossil fuels at current rates. Since physical systems tend toward entropy or disorder (i.e., molecules will tend to spread out such that energy needs to be expended to contain them), keeping carbon in the ground as opposed to attempting to remove it from the ambient air after the fact is inherently more efficient (cf. Huesemann and Huesemann 2011, 79–82). CDR techniques, even with assuming full actualization in terms of feasibility and scalability, are not tantamount to emissions reductions and are limited in what they can achieve.

The NASEM report on NETs takes an instrumental approach, arguing:

Combustion of 1 gallon of gasoline releases approximately 10 kg of CO_2 to the atmosphere. Capturing 10 kg of CO_2 from the atmosphere and permanently sequestering it therefore has the same effect on atmospheric CO_2 as any mitigation method that simultaneously prevents combustion of 1 gallon of gasoline. (NASEM 2019, 23)

However, for this to be the case would require also capturing or accounting for all the energy (not to mention other environmental harms) used to capture that first 10 kilograms. In terms of energy and impact, capturing from the ambient air is inherently inefficient compared to not polluting in the first place. Even if clean energy is used, in the case of DAC, to capture and sequester the CO_2, that clean energy could have been repurposed elsewhere. The Committee concedes this point in their final chapter saying, "Gt-scale direct air capture thus necessitates an enormous increase in low- or zero-carbon energy to meet these energy demands, which would compete with use of such energy sources to mitigate emissions from other sectors" (NASEM 2019, 365).

The IPCC points out that in the quest for net-zero (or net negative) emissions, it cannot be presumed that CDR necessarily constitutes a one for one tradeoff toward that goal in terms of measurement by metric tons of GHGs emitted versus captured: "The atmospheric CO_2 decrease from anthropogenic CO_2 removals could be up to 10% less than the atmospheric CO_2 increase from an equal amount of CO_2 emissions, depending on the total amount of CDR" (IPCC 2018, 39). Moreover, CDR is limited in changing the course of climatic impacts already occurring. According to the IPCC:

If global net negative CO_2 emissions were to be achieved and be sustained, the global CO_2-induced surface temperature increase would be gradually reversed but other climate changes would continue in their current direction for decades to millennia (high confidence). For instance, it would take several centuries to millennia for global mean sea level to reverse course even under large net negative CO_2 emissions (high confidence). (IPCC 2018, 39)

Climate models also indicate that even with "massive CDR interventions," CDR cannot "restore pre-industrial conditions in the ocean by reducing the atmospheric CO_2 concentration back to its pre-industrial level" (Mathesius et al. 2015). As stated by Earth system scientist Steven Davis, "We can't think that we should be able to make up for poor decisions today by buying negative emissions later. [. . .] It's going to be a lot easier, cheaper and less risky to just tackle these emissions before they're in the atmosphere" (quoted in Upton 2015). Similarly, with emphasis on the common understanding that CDR will be evaluated on costs, Douglas MacMartin states in his Congressional testimony on geoengineering that air capture technology "is almost certain to be technically feasible, but right now probably too expensive. It's almost certain to be cheaper to not put it in in the first place than to take it out after you've put it in" (Congressional Hearing 2017b).

In terms of processing carbon dioxide in the atmosphere, there is a certain irony (or, to put it more strongly, irrationality) in the attempt to invent machines that can absorb carbon dioxide contemporaneously to ongoing rates of deforestation, especially of tropical forests which are primary to the natural processing of CO_2. The replacement of natural systems with artificial carbon systems is analogous to the idea of "artificial life on a dead planet" in which "Life" is "radically abstracted, resituated, and reconfigured" as "broader nature" is devalued and "becomes a sink for pollution and other 'externalities'" of production (Thorpe 2016, 67, 80; see also Thorpe 2013; Thorpe and Jacobson 2021). So long as deforestation of older forests subsists concurrently, even the relatively benign and "sensible" land management options categorized as CDR[5] like reforestation and afforestation contend with paradoxes of entropy (the move toward disorder as carbon stored in mature forests is released into the atmosphere) and latency (delayed effectiveness as a carbon sink since young forests cannot absorb as much CO_2 as mature forests). Furthermore, "fragmentation of tropical forests is likely to increase emissions of CO_2 and other greenhouse gases above and beyond that caused by deforestation per se" due to ecological changes to remaining forest stands that lead to increased tree mortality and higher risks of forest fires (Laurance, Laurance and Delamonica 1998; Laurance, Vasconcelos and Lovejoy 2000). Just like keeping carbon in the ground is more efficient than removing it after the fact, keeping mature forests intact is more effective than reforestation (the latter of which, incidentally, is counted among CDR strategies).

Paralleling the physical entropy of redistributing and dispersing carbon into the ambient air is the social redistribution of the burden for that carbon pollution. Much debated policy proposals for pricing carbon emissions would place a cost on emissions incurred by the producer (and ultimately consumer) of those emissions. In contrast, CDR is modeled after the tradition of public appropriation of externalities wherein private entities profit from production, but the broader public incurs the environmental costs of that production. In the case of CDR, this dynamic is rather straightforward as private industry would sell CDR services to governments, internalizing the profit of removing excess carbon (caused in the first place primarily by private industry), ultimately paid for by the general populations of the participating states.

Ethicist Stephen Gardiner explains that risks of climate change are both "spatially and temporally dispersed to different countries and generations" (Gardiner 2011, 172). The same can be said of the costs and risks of trying to address climate change through geoengineering. Spatially, in terms of international geopolitics, Richard York adds that CDR techniques like BECCS are "inherently" imperialistic, as such projects would "require controlling land in many nations around the world, harking back to earlier colonial projects," with affluent countries that are primarily responsible for climate change in the first place asserting their vision of mitigation through CDR projects upon the land of less developed countries (York 2021, 185). Temporally, future generations pay the costs of past generations' pollution, which also brings up issues of intergenerational ethics in regard to who internalizes the costs, risks, and harms associated with geoengineering projects (Gardiner 2011).

Geoengineering has often been described as a "Plan B" to mitigation. While CDR has come to be considered a more immediate option to pursue alongside other mitigation efforts, albedo modification is still most often discussed in terms of a Plan B scenario (as will be discussed in chapter 2). However, this categorization can be misleading since Plan B presumes a Plan A, in this case mitigation; yet the Plan A of mitigation has not been seriously and fully pursued due to obstruction by powerful economic and political interests (cf. Schneider and Fuhr 2021, 51).

Furthermore, as a Plan B, it is an illusory alternative since it is encumbered with similar political challenges to those that have hampered mitigation efforts to date. As geoengineering research advocate, the cosmologist and astrophysicist Lord Martin Rees (who was president of the Royal Society between 2005 and 2010) recognizes:

Geoengineering would be an utter political nightmare: not all nations would want to adjust the thermostat the same way. There could be unintended side-effects. Regional weather patterns may change. Moreover, the warming would return with a vengeance if the countermeasures were ever discontinued; and other consequences of rising CO_2 (especially the deleterious effects of ocean acidification) would be unchecked. (Rees 2013)

As he indicates in terms of regional effects and potential arguments over setting "the thermostat," various regions of the globe are likely to be differentially affected should albedo modification be pursued. In the execution of geoengineering, there are likely to be winners and losers in terms of climatic effects. For example, models have indicated that global albedo modification may significantly impact the reliability of Asian and African monsoons on which those regions' agriculture depends (e.g., Goodell 2010, 9, 132; Hamilton 2013a, 64; National Research Council 2015b, 46; Da-Allada et al. 2020). Irrespective of specific climatic interests, there will be individuals, networks, and nations opposed on principle (e.g., ETC Group 2010).

Given the globality of proposals as well as the extent of risks and concerns, it can be reasonably presumed that the same techno-political barriers that have obstructed international mitigation efforts would also hamper implementing geoengineering in a multilateral fashion, as advocates insist would need to be the case for geoengineering to be "legitimate." Legitimate pursuit of geoengineering is indicated as requiring "international" collaboration and cooperation (e.g., Royal Society 2009; National Research Council 2015b; NASEM 2021, 6–13). Furthermore, just as would be necessary for mitigation, careful monitoring and regulation of GHG emissions would be required in order to pursue climate engineering based on real-world conditions. Along these lines, Andreas Malm argues that any large-scale geoengineering venture, whether SRM or CDR, requires central planning at an equal or greater level than that which has been effectively ruled out in terms of controlling emissions (Malm 2021). In short, geoengineering poses new challenges of international cooperation and coordination without solving old ones.

Another common concern regarding geoengineering is moral hazard, which is "a term derived from insurance, and arises where a newly-insured party is more inclined to undertake risky behaviour than previously because compensation is available" (Royal Society 2009, 37). The Royal Society describes the risk as follows: "In the context of geoengineering, the risk is that major efforts in geoengineering may lead to a reduction of effort in mitigation and/or adaptation because of a premature conviction that geoengineering has provided 'insurance' against climate change" (Royal Society 2009, 37). Similarly, as articulated by the NRC: "There is a risk that research on albedo modification could distract from efforts to mitigate greenhouse gas emissions" (National Research Council 2015b, 8, cf. 6, 147).[6] Both influential organizations writing on the matter, however, ultimately dismiss the moral hazard concern in terms of pursuing geoengineering research. The NRC concludes that "as a society, we have reached a point where the severity of the potential risks from climate change appears to outweigh the potential risks from the moral hazard associated with a suitably designed and governed research program" (National Research Council 2015b, 8). The Royal Society further poses a counterargument to the moral hazard concern, contending

> there is little empirical evidence to support or refute the moral hazard argument in relation to geoengineering, (although there has been little research in this area), and it is possible that geoengineering actions could galvanise people into demanding more effective mitigation action. (Royal Society 2009, 39)

Despite the dismissal of moral hazard concerns on the part of these major research institutions, such concerns are consistent with the broad consensus that geoengineering, particularly albedo modification, is deeply fraught with material, ecological, and social risks. As phrased by the NRC, "Proposed albedo modification approaches introduce environmental, ethical, social, political, economic, and legal risks associated with intended and unintended consequences" (National Research Council 2015b, 5). Furthermore, since albedo modification would create completely novel climatic conditions, risks are joined by unknowns. Again, as stated by the NRC, "Intervening in the climate system through albedo modification therefore does not constitute an 'undoing' of the effects of increased CO_2 but rather a potential means of damage reduction that entails novel and partly unknown risks and outcomes" (National Research Council 2015b, 35). In regard to these novel circumstances, a caveat is made regarding the limitations of knowledge inherent in the proposed research, including models: "This real system has far greater complexity than does any model, and thus no model of this system can provide a quantitatively reliable detailed prediction of how Earth will respond to a novel occurrence" (National Research Council 2015b, 39). There is general agreement that "Albedo modification presents a number of risks and expected repercussions" (National Research Council 2015b, 6). These include ecological harms, changes to the hydrological system, changes to precipitation patterns, and potential effects on agriculture and human health (e.g., National Research Council 2015b, 6, 28, 33–35, 46, 53).

The unique scale and intentionality of environmental effects constituted by geoengineering makes it a particularly compelling site to study modernity and global risk, technology and society, and how technological expertise is presented and received. While other authors have written insightfully about social dynamics and critiques of geoengineering more generally (e.g., Baskin 2019; Stilgoe 2015; Hulme 2014; Hamilton 2013a; Goodell 2010; Kintisch 2010; Parkinson 2010; Saspinski, Buck and Malm 2021), the empirical portion of this manuscript is focused on discourse analysis of geoengineering as a method to understand the mainstreaming process that has been so pronounced in this emerging technology's trajectory.

OVERVIEW AND PREVIEW

The scale of risks associated with both climate change in the first place and geoengineering proposals in response to it are unique in their scale of global totality. Both are also fraught with complexities of interacting environmental and technological risks, international and national politics, economic interests, and public perceptions. While distinct in scale and technological novelty, the emergence and mainstreaming of geoengineering fits within a broader social and historical dynamic between modernity, risk, and the environment. Chapter 1 situates the case of geoengineering within the broader picture of risk, modernity, and climate politics.

Subsequently, the following three sections present the core empirical results of the discourse analysis of geoengineering. Ian Welsh argues that science cannot be understood as a "unified set of institutions, practices and techniques," but rather "it is crucially important to pay attention to the particular discourses which are constructed around particular technologies. The extent to which a technological narrative articulates sympathetically with other ascendant discourses plays a crucial role in determining its success" (Welsh 2000, 4). Extrapolating Welsh's assertion, which he made in regard to the case study of nuclear technology, an individual field of science such as geoengineering can be understood as constituting "a particular scientific social movement seeking to transform society through the acceptance of particular sets of knowledge claims and acceptance of the associated social and technical practices" (Welsh 2000, 5). Analyzing discourse facilitates better understanding of such knowledge claims, the intersections of ideology and science, and the diffusion of ideas between technical experts and the public.

Discourse around a topic like geoengineering is particularly salient. Other than a limited number of experiments to date, the concept is still manifested primarily within discourse and imaginaries (Stilgoe 2015; Markusson 2013). Moreover, the framing and conceptualization of geoengineering are subject to asymmetrical power relationships and knowledge claims. It has been noted that the framing of emerging or "upstream" technologies like geoengineering can significantly affect their reception among the public (Luokkanen, Huttunen and Hildén 2014; Bellamy 2013; Bellamy et al. 2012). Hence, discourse is central to the politics of geoengineering.

In particular, examination of the relevant discourse on geoengineering within and between the scientific, political, and public spheres contributes to understanding its shifting role in society and politics as well as the material interests and ideological beliefs underlying the dynamic unfolding process. The public nature of this issue lends itself well to content analysis, integrating principles of historical analysis and critical discourse analysis to examine the documents that both reflect and drive the trajectory of geoengineering, including influential science policy reports (especially the Royal Society, National Academies, and NRC reports), journalistic articles (news reports, expository reports, editorial commentary), as well as written records and video-recordings of U.S. Congressional hearings on the subject.

Discourse analysis is used to examine framing, staging, narrative construction, and argumentation within texts and how certain narratives "come to prevail" and, hence, influence policy and outcomes (Fairclough and Fairclough 2012, 1, 6). The approach to discourse analysis here is based upon the constitutive theory of language and discourse, which understands language, its conventions, and uses as not only expressing thoughts and ideas about objects but also affecting, redefining, and constructing such objects and their related contexts (see Mehan, Nathanson and Skelly 1990, 135; Foucault 2010). In this way, discourse embodies an intersection of communication, understanding, and action. Discourse not only guides decision-making and material outcomes but is also itself shaped by social and political interests. As such, this study of geoengineering discourse and the competing narratives within it draws upon the "politics of representation" and its constitutive approach of understanding discourse as influencing "ways of thinking and ways of acting" as well as highlighting the "competition over meaning" in which "proponents of various positions [. . .] attempt to capture or dominate certain modes of representation" (Mehan, Nathanson and Skelly 1990, 137; Mehan and Wills 1988, 364). Political discourse analysis

> is based on a view of politics in which the concepts of deliberation and decision-making in contexts of uncertainty, risk and persistant disagreement are central. This is a view of politics in which the question of *action*, or *what to do*, is the fundamental question. (Fairclough and Fairclough 2012, 17)

Language affects outcomes and "political struggles have always been partly struggles over the dominant language" (Fairclough 2000, 3).

Due to the scale and risk, geoengineering inherently involves science crossing over into international law and policy. Section II (chapters 2 and 3) explores the intersection of science and policy with respect to geoengineering through discourse analysis of key scientific reports that are designed to inform policy debates on the potential strategies of countering climate change through climate engineering. It focuses on several of the most important and influential geoengineering science policy reports published: The

Royal Society's 2009 (UK) geoengineering report, the NRC of the National Academy of Sciences two-volume report in 2015 (United States), and the NASEM 2021 (United States) Solar Geoengineering report, with additional consideration of other related reports by these scientific bodies as well as the IPCC. These reports construct conceptualizations of geoengineering and influence its trajectory of research and development due to their central positions in informing policy makers, news media, and, in turn, public perception. They represent a manifestation of climate engineering's shift toward mainstream consideration and serve an important role in setting the discursive tone for discussion of the climate crisis and consideration of geoengineering. As will be discussed, these reports employ certain discursive strategies that effectively construct notions of legitimacy and normalcy related to geoengineering.

While science policy reports represent the translation of geoengineering imaginaries into language accessible to political decision-makers, news media bring these concepts to a broad audience. Section III (chapters 4 and 5) examines discourse in the public sphere primarily through analyzing trends present in English-language newspaper and magazine articles focused on geoengineering, with a focus on U.S. and UK publications. In spite of a broad range of voices, vantage points, opinions, framings, and highlighted facts represented in this corpus, the analysis identifies discursive themes that become conventionalized in public discourse on geoengineering. These conventions uphold certain narratives that construct public conceptualization of the "sociotechnical imaginaries" involved in geoengineering (Jasanoff 2015; Bellamy et al. 2012; Stilgoe 2016; Healey 2014; Markusson 2013; Corner et al. 2013).

Section IV (chapter 6) focuses on contentious political discourse and expert testimony within Congressional hearings. Within the confines of these generally polite and formal proceedings, there are two elite classes of actors present and asserting authority: expert witnesses and Congressional representatives. These hearings provide insight into geoengineering deliberations in which discursive participants are co-present and interacting in live time. Moreover, the interactions also demonstrate engagement with external material and context that gets drawn into the hearings despite the clear and specific purposes delineated within each hearing charter. Particularly, in the case of geoengineering hearings, the politicized contention over climate change gets interwoven into the discussion along with politicians contending for legitimacy in the staging of broad ideological values such as environmental protection, economic development, and the appropriate scope of government regulation. The U.S. House of Representatives has conducted four hearings on geoengineering from 2009 through 2016. The analysis examines the dramaturgical performances within these hearings and analyzes which elements of discursive enactment remain consistent over time and what changes occur over the seven-year time period.

The final section (conclusion) returns to the theoretical themes raised in the introduction and considers the place of geoengineering in relation to world risk society, reflexive modernity, life politics, and the politics of unsustainability. As an alternative to the original "Plan A" of emissions reductions, geoengineering provides a particularly salient lens through which to analyze these issues. It not only addresses global risks but also brings its own risks, known and unknown. It inherently involves a great extent of human knowledge and ingenuity but problematizes the complex questions of social reflexivity in the face of the global risks of modernity. It also adds new dimensions to Blühdorn's concepts of post-ecologism, simulative politics, and the politics of unsustainability both in its own right as a contested field and also in its positioning in relation to other climate solutions. The opposite of an instrumental and technocratic approach can be seen in the life politics and values promoted by environmental social movements, as will be discussed in the final chapter.

NOTES

1. For example, the 1993 publication in *Climate Change* that suggested "the effects of global warming could be countered by increasing the radius of the Earth's orbit around the Sun" (Hamilton 2013, 3).

2. Exemplifying collaboration between engineers, scientists and environmentalists on the topic of geoengineering, the Solar Radiation Management Governance Initiative's "Solar Radiation Management" (2011) report was jointly published by the Environmental Defense Fund (EDF), The Royal Society, and TWAS (the academy for sciences for the developing world).

3. The term "solar radiation management" implies that it may be feasible, necessary and/or appropriate to subject the sun's rays to human management. This implication of beneficial human management of solar radiation is problematic considering that albedo modification proposals are prima facie indicative of a failure of human management of the atmosphere.

4. To give a sense of this, the acronym "CDR" appears 303 times and the terms "net negative" and/or "negative emissions" appear 126 times in the IPCC Working Group I's "Climate Change 2021: The Physical Science Basis" full report ("Final Government Distribution" version downloaded October 14, 2021 from the IPCC website).

5. Variants of the word "sensible" are often used in reference to reforestation and afforestation CDR approaches (e.g., National Research Council 2015a, 88; Mathiesen 2015).

6. Some scholars have rebranded the term "moral hazard" as related to albedo modification research to the more specific and descriptive "mitigation obstruction" (Halstead 2018; Morrow 2014).

Chapter 1

Risk, Climate Politics, and the Challenge of Reflexive Modernization

MODERNITY, RISK, AND CLIMATE POLITICS

Late modern society is characterized by awareness of risk and uncertainty as key aspects of the human relationship with the natural environment on which we depend. Some scholars argue that the risk we face is unprecedented in its scale and magnitude. At the beginning of the 1970s, when the environmental movement was becoming increasingly socially and politically influential, one of its key intellectual leaders, the biologist Barry Commoner, stated the dangers posed by environmental degradation starkly:

> To survive on the earth, human beings require the stable, continuing existence of a suitable environment. Yet the evidence is overwhelming that the way in which we now live on the earth is driving its thin, life-supporting skin, and ourselves with it, to destruction. (1971, 14)

More recently, Anthony Giddens has argued that global climate change is unlike any other challenge we have faced before due to the enormity of its potential disaster and its global scale (2009, 2). Similarly, according to Ulrich Beck, in modern society, we are faced with risk of "self-destruction of all life on earth due to human interventions" (Beck 2009, 27).

For the last century, modern society has tested the boundaries of risks from chemical toxins, industrial pollutants, nuclear reactors, deforestation, and mineral extraction, including new forms of fossil fuel extraction such as deepwater drilling and chemical fracking. The ecological ramifications include localized and globalized pollution, climate change, glacial and icecap melting, sea rise, ocean acidification, and loss of biodiversity with "exceptionally high" rates of extinction (Pimm et al. 2014; IPCC 2014; Ceballos, Ehrlich and

Dirzo 2017). The prospect of geoengineering adds a new dimension to risk since its environmental impacts would include *intentionally and reflexively instituted risks* in addition to the types of incidental environmental risks to which society has become accustomed.

Giddens argues that modern societies have "broken away from nature" (Giddens 1990, 63). He contends that "what used to be natural is now either the product of, or influenced by, human activity" (Giddens 1998, 58). However, despite traditional notions of nature changing or disappearing, nature arguably remains an important conceptual category (Dickens 1999, 102–104; Thorpe and Jacobson 2013, 117). Giddens considers the human relationship with nature as a critical locus of ontological insecurity in modernity (see Thorpe and Jacobson 2013, 103–105, 118; cf. Beck 2009, 121). Ontological security means the ability to remove from consciousness the existential dilemma of "how should we live?," relying on received authoritative frameworks for making sense of, and establishing boundaries around, everyday life. Initially in modernity (or what Giddens calls "simple modernity"), science and technology replaced tradition as the primary basis of ontological security (see Thorpe and Jacobson 2013, 103–105, 118; cf. Beck 2009, 121). The ability to control nature and thereby force it into the background of quotidian social life has been an important modern dimension of ontological security. However, as modern societies come to be faced with the destructive ecological consequences of modern technological means of controlling nature, risk becomes reflexive (as our very attempts to contain risk produce new risks) and modern societies are forced into a state of reflexivity with regard to risk (recognizing the potentially unruly unintended consequences of the modern attempt to control nature and no longer able to bracket risk from everyday life) (see Beck, Giddens and Lash 1994).

Climate change is a key (arguably, the most important) instantiation of this reflexivity of risk. Climate change problematizes the conception of external nature as intrinsically "natural" elements such as weather become affected by human intervention. This heightens "the ontological insecurity of high modernity" as "ecological problems such as climate change re-open the existential contradiction under conditions in which this is no longer adequately mediated by social institutions" (Thorpe and Jacobson 2013, 117–118). While the distinction becomes blurred, nature does not disappear or become entirely submerged into society. Rather, as Andreas Malm argues, "climate change sweeps back and forth between the two regions traditionally referred to as 'nature' and 'society'" (Malm 2018, 15).

Ulrich Beck, whose distinguished career centered around theorizing the social components of risk, argues that while humanity has always faced risk, what distinguishes modern risks beyond their global scale and magnitude is the means of their creation. Within the paradigm of risk society, corresponding with late modernity, we are "concerned no longer exclusively with making nature useful, or with releasing mankind from traditional constraints, but also and essentially with problems resulting from techno-economic development itself" (Beck 1992, 19). The risks we face today "are a product of human hands and minds, of the link between technical knowledge and the economic utility calculus" (2009, 25). While some risks are timeless, modern risks are distinct in being generated by "conscious decisions" and "calculations for which hazards represent the inevitable downside of progress" (Beck 2009, 25). Risk has become interwoven into modern society and economy: "In advanced modernity the social production of *wealth* is systematically accompanied by the social production of *risks*" (Beck 1992, 19, emphases in original). Hence, Beck proposes that modern society ails from its "triumphs" (rather than its defeats) because its successes bring with them new risks (2009, 8, 22–23, 30).

Beck's theory of the risk society as a new formation replacing an earlier form of modernity draws, to a certain extent, on predictions of a post-industrial society in the 1960s and 1970s. Post-industrial society theorist Daniel Bell had observed that technological advancements can have "deleterious side effects, with second-order and third-order consequences that are often overlooked and certainly unintended" (Bell 1973, 26–27). According to Beck, "The gain in power from techno-economic 'progress' is being increasingly overshadowed by the production of risks," which have expanded from "latent side effects" to global issues of "central importance" (Beck 1992, 13). Climate change is a clear example of such a modern global risk resulting directly from economic progress. Climate change "is a product of successful industrialization," one that constitutes a new global risk that is delocalized, incalculable and non-compensable (Beck 2009, 8, 52). "Risks can no longer be dismissed as side effects" of modernization (Beck 2009, 194).

While society has created these new global risks, conversely and reflexively, people are in turn affected by these environmental risks on multiple levels beyond the direct material threats. Giddens and other social theorists have recognized that the human relationship with nature represents a critical component of ontological insecurity in modernity (Giddens 1991; Giddens 1981; Thorpe and Jacobson 2013, 103–105, 118; Norgaard 2006, 380). Giddens theorized that first modernity facilitated the suppression of existential dilemmas through replacing existential contradictions with social relationships and structural contradictions lubricated by the routinization and predictability of everyday life (Giddens 1981). Security came to depend on trust in "abstract systems" of technology and scientific expertise (Giddens 1990, 92–93, 112–113; Giddens 1994, 80). However, existential crisis reemerges as global risks such as climate change problematize the sense of security and the sequestration of risk.

The existential crisis, raised by climate change and other global risks, does not occur just despite the efforts of modernity but rather as a by-product of modernity (Beck 2009). Giddens argues that "in conditions of high modernity, crisis becomes normalised" (Giddens 1991, 184). "The crisis-prone nature of late modernity thus has unsettling consequences" for society and individuals both in terms of fueling uncertainty and threatening "the very core of self-identity" (Giddens 1991, 184–185). There is a central need for ontological security, meaning "the confidence that most human beings have in the continuity of their self-identity and in the constancy of the surrounding social and material environments of action" (Giddens 1990, 92). The "transformation of nature introduces a new kind of ontological insecurity" with "global warming as an example of this new unpredictability arising not from brute nature but from the unintended consequences of industrial society" (Thorpe and Jacobson 2013, 105).

Ecological politics are demonstrative of "reflexive modernization" (Giddens 1994; Beck, Giddens and Lash 1994; Beck 1992; Giddens 1990; Thorpe and Jacobson 2013). Furthermore:

> The emergence of ecological politics demonstrates that modernity is no longer able to bracket the existential contradiction rooted in human beings' relationship to nature. A key dimension of the reflexivity of high modernity is that we can no longer treat the problem of nature as progressively solved through instrumental control. Instead, how we, as conscious agents, relate to nature becomes again a problem of morality and meaning as well as of scientific understanding. (Thorpe and Jacobson 2013, 104–105)

Beck indicates that modern reflexive risks are the objects of new forms of political action, which he calls *subpolitics*, such as environmentalism, which operate outside and, in a certain sense, beneath the activity of the centralized institutions of representation and government (Beck 1997). Relatedly, Giddens's concept of *life politics* encapsulates "disputes and struggles about how (as individuals and as collective humanity) we should live in a world where what used to be fixed either by nature or tradition is now subject to human decisions" (Giddens 1994, 14–15). However, in regard to climate change, the central ecological crisis of modernity, Giddens retreats from engagement with life politics (Giddens 2009; Thorpe and Jacobson 2013). The extent to which climate change resurfaces ontological insecurities complicates its relationship to life politics.

While environmental activism and life politics remain centrally important to climate politics, there has also been a trend of "collective avoiding" (Norgaard 2006). This means that, despite knowledge and concern regarding climate change, people avoid thinking about its catastrophic risks in order to protect themselves from the psychological toll of unpleasant and troubling emotions involved in grappling with existential crisis (Norgaard 2006). Climate change specifically has the potential to fundamentally alter the material basis of human society in dramatic and intractable ways. "At the deepest level," according to Norgaard, it threatens "people's sense of the continuity of life" (2006, 380), posing a potential existential crisis within modern society. People cope with these risks in various ways. One coping mechanism is suppression or avoidance. Norgaard's (2006) study of informed and educated Norwegians found that people avoid thinking about the catastrophic risks of climate change despite, or perhaps because of, belief in its actuality and knowledge of its risks. She calls this process the "social organization of denial," which is a form of collective emotional avoidance despite intellectual understanding. This form of denial helps toward explaining the relative lack of action on the issue of climate change despite knowledge and concern over it.

A contrasting form of suppression is that which occurs for the sake of self-interest rather than emotional self-preservation. Climate change denial and so-called skepticism fit in this category. As discussed previously, Oreskes and Conway (2010) trace the historical circumstances that have contributed to the downplaying of global warming. Groups with a denial agenda were able to cause public confusion and doubt regarding the level of scientific consensus on climate change through steering public discourse on the issue. This was especially facilitated through the presentation of a two-sided "debate" within mass media that displayed climate change as an issue of scientific controversy, thereby distorting the impression of climate change disproportionately in favor of skeptics and deniers (Oreskes and Conway 2010, 214; cf. Perrow 2011, xxxviii). The subsequently entrenched questioning and distortion of climate science have contributed toward maintaining inertia and lack of political will to act on the crisis of climate change.

At an institutional level, two primary challenges have stood in the way of effective climate politics. First is the need for global cooperation and the conflicting political-economic interests that complicate it. According to Beck, global risks can only be addressed through international cooperation fostered by the "cosmopolitan moment" that such risks present (2009, 15, 55–56). The very nature of global risks, and the mutual sense of vulnerability they present, has potential to "open up a moral and political space that can give rise to a civil culture of responsibility that transcends borders and conflicts" (Beck 2009, 57). He points to *cosmopolitan realpolitik* as the hybrid fusion of state interests and the cosmopolitan moment, which compels the inclusion of cultural others, as the political tool for addressing global risks (Beck 2009, 66, 56). Of course, the need for international cooperation on global environmental risks is clear and frequently articulated. Giddens argues, in

The Politics of Climate Change, that "effective response must involve nations working together, even countries whose interests in other respects might seem opposed" (Giddens 2009, 14). However, global collaboration has been challenged to date by competition between nation-states (Harris 2013; Victor 2011) and other economic interests invested in the existing energy economy.

The other core challenge is the scope of changes to the world economic system necessary to effectively mitigate climate change. This system is defined by an emphasis on indefinite growth and has been characterized by environmental sociologists as "the treadmill of production" (Gould, Pellow and Schnaiberg 2008). Constant growth is a central component of the economic system and is necessary to avert *economic crises* (Harvey 2010; Harman 2010) while simultaneously feeding *environmental crises*. In contrast to economic activities characterized by growth, the ecosphere is characterized by cyclical processes, balance, and some intrinsic limits to growth: "One can argue whether the ecosphere, in its pre-human, natural condition or in its present one, operates near its intrinsic limit; but that there is some limit, and that the system's operation does not permit indefinitely continued growth, is undeniable" (Commoner 1971, 120–121). Commoner calls this the

> fundamental paradox of man's life on the earth: that human civilization involves a series of cyclically interdependent processes, most of which have a built-in tendency to grow, except one—the natural, irreplaceable, absolutely essential resources represented by the earth's minerals and the ecosphere. (1971, 122–123)

Similarly, treadmill of production theory points to economic change as the primary driving factor in escalating environmental impacts. Following World War II, capital investment changed quantitatively and qualitatively as more capital was accumulated, and it was increasingly allocated to "replacing production labor with new technologies to increase profits" (Gould, Pellow and Schnaiberg 2008, 7; cf. Commoner 1971, 177). These new technologies were more environmentally disruptive, requiring greater energy and chemical inputs (Gould, Pellow and Schnaiberg 2008, 7). Since the investments constituted sunk capital requiring sustained or increased production levels to justify, "capital investment led to greater demand for natural resources, for a given level of social welfare" (Gould, Pellow and Schnaiberg 2008, 11). Profits came to be "increasingly invested in new technologies rather than in expanding employment or raising the status of workers" (Gould, Pellow and Schnaiberg 2008, xii).

This process was self-perpetuating as investment necessitated further investment: "Each round of socially dislocating growth generated increased, rather than decreased, social support for allocating investment to accelerating the treadmill of production" (Gould, Pellow and Schnaiberg 2008, 12). A central trend of "modern technology has been to displace human labor in goods production and substitute physical capital and inanimate energy supplies" resulting in increased production and profits as well as increased environmental impact and displacement of workers (Schnaiberg 1980, 415–416). As technology displaced labor, it came to be thought that further growth was the only possible course for maintaining employment levels. Hence, ironically, "economic growth was viewed as the primary solution to the negative social impacts of economic growth" (Gould, Pellow and Schnaiberg 2008, 12). According to this theory, social progress is thwarted as society remains on the economic treadmill, figuratively "running in place," as social efficiency is decreased, resulting in "increased rates of ecosystem depletion (resource extraction) and ecosystem pollution (dumping of wastes into ecosystems)" (Gould, Pellow and Schnaiberg 2008, 12).

Like treadmill theorists, Commoner (1971) made a compelling case for the predominance of technology in affecting the major qualitative and quantitative shifts of environmental impact since World War II.[1] According to Commoner, the environmental crisis, at least in the United States, has been largely tied to the transformation of productive technology in which "productive technologies with intense impacts on the environment have displaced less destructive ones. The environmental crisis is the inevitable result of this counterecological pattern of growth" (Commoner 1971, 177). Similarly, disputing the overemphasis on population rather than production, Allan Schnaiberg concludes that "production, rather than reproduction, is the crucial factor in biospheric disorganization" (1980, 98). In the 25 years following World War II, Commoner points out that the United States achieved not only a 126% rise in GDP but also a rise in environmental pollution that grew at ten times that rate (1971, 146). The difference is explained by externalities.

Externalities refer to the costs of production that are not internalized by the firms involved, including environmental or social consequences borne by the public. The internalization of benefits (profit), with the externalizing of costs (social dislocation and environmental degradation), drives the tendency to pursue economic advancement at the cost of society: "Ecological irresponsibility can pay—for the entrepreneur, but not for society as a whole" (Commoner 1971, 267). As Giddens states, the "environmental costs entailed by economic processes often form what economists call 'externalities'—they are not paid for by those who incur them" (2009, 5). For example, with the absence of a pricing scheme for carbon emissions, the release of GHGs is an externality. Firms get to internalize the profits incurred through production, but they externalize the environmental costs, such as carbon pollution, which affect the broader public and the environment. Charles Perrow argues that the free-market capitalism promoted since the Reagan/Thatcher era has exacerbated the problem of externalities and contributed to "global warming at an increasing rate" (Perrow 2011, xxix).

Externalities are often considered a market failure, which could be corrected with appropriate policy. Giddens argues that the "aim of public policy should be to make sure that, wherever possible, such costs are internalized—that is, brought into the marketplace" (2009, 5). This objective, however, is complicated by technical considerations of how to measure and price pollutants as well as knowledge asymmetries favoring business, power dynamics, and political interests. Perrow argues for public policy that includes regulation as well as decentralization of power and a reduction in the scale of high-risk operations (Perrow 2011, 35–36, 318–321). Regulation of industries, with jurisdictions free to increase (but not decrease) standards, is a more direct method of limiting the consequences of externalities than through pricing mechanisms. Often, however, proposed solutions to the externalities problem include a combination of regulation and pricing mechanisms.

Foster challenges the notion of internalizing externalities into the market. He identifies the standard position of environmental economics as aimed "at the creation of markets to solve problems of pollution and environmental degradation" (Foster 2002, 29). This can be pursued through "imposition of taxes or subsidies that will increase the costs of inflicting environmental damage and the benefits of environmental improvements" or through the imposition of new markets "such as tradable pollution permits" (2002, 29–30). However, Foster is skeptical of both approaches due to their strategy of "turning the environment into a set of commodities" and "overcoming the so-called market failures of the environment by constructing replacement markets for environmental products" (2002, 30). "Nature," he argues, "is not a commodity produced to be sold on the market according to economic laws of supply and demand" (2002, 30).

This argument is reminiscent of Karl Polanyi's *The Great Transformation*, from a generation earlier, where he argued that land, along with labor and money, constitutes a "fictitious commodity" (Polanyi [1944] 2001). Land, labor and money are essential to the functioning of markets, but they are not commodities created for the sake of exchange. Polanyi warns that including land (i.e., nature) and labor (i.e., people) "in the market mechanism means to subordinate the substance of society itself to the laws of the market" ([1944] 2001, 75). Marion Fourcade's (2011) contemporary analysis of valuation of nature in response to oil spills speaks to the complexity of placing pecuniary value onto elements of nature.

Foster further argues that ecological sustainability

> can be undermined not only through the economy failing to take environmental costs into account (the externalization of costs to the environment), as is

commonly supposed, but also by the attempted incorporation of the environment into the economy—the commodification of nature. (Foster 2002, 30)

By this logic, there is a flipside to externalities that can be traced to commodification of nature, which also promotes ecological degradation and exploitation. According to Foster, this can be more pernicious than the problem of externalities:

> It is not so much the failure to internalize large parts of nature into the economy that is the source of environmental problems, but rather that more and more of nature is reduced to mere cash nexus and is not treated in accordance with broader, more ecological principles. (Foster 2002, 33)

Similarly, John Urry finds that "as a consequence of neo-liberalization, sustainability has often been reformulated as 'sustainable development,' with an astonishing array of industries and developments seen as contributing to what is 'sustainable'" (Urry 2011, 27).

While economic and technological changes have been important factors in modern production trends and the correlating environmental effects, these changes do not exist in a vacuum. For Schnaiberg, the treadmill of production is driven by institutional factors, including social structures and class dynamics. The constituencies of capital, labor, and the state each have their own interests in regard to production expansion (Schnaiberg 1980, 211). He refers to these three influential social sectors as the *growth coalition*. Ultimately, then the treadmill is socially constructed but with substantial environmental consequences:

> The logic of the treadmill is that of an ever-growing need for capital investment in order to generate a given volume of social welfare. [. . .] From the environment, it requires growing inputs of energy and material to create a given level of socioeconomic welfare. When resources are constrained, the treadmill searches for alternative sources rather than conserving and restructuring production. The treadmill operates in this way to maintain its profits and its social control over production. (Schnaiberg 1980, 418)

The perpetuation is geared by social structures and the interests of various groups with economic power and political influence.

Due to the inherent dynamic of capitalist economics and the environment, Foster argues against those who see capitalism as containing "within itself the solution to global environmental problems" (2002, 22). A key example of attempts to handle ecological problems within the framework of capitalist relations and institutions is Giddens's *The Politics of Climate Change* (2009). Here, Giddens suggests that "more of the same will be needed, not less, if we are seriously to confront the problems of climate change" (2009, 6). In this sense, Giddens suggests that increased technical innovation is what is necessary to solve the global environmental crisis of climate change.

His argument is similar to those put forth by proponents of ecological modernization theory (EMT) in suggesting that environmental issues could be best addressed "by drawing them into the existing framework of social economic institutions, rather than contesting those institutions" (Giddens 2009, 70). EMT stands in stark contrast to most social analyses of climate change in its optimism that sustainability can arise "from a combination of economic development, technological innovation, and institutional reform" (Thorpe and Jacobson 2013, 112; Dryzek 2005, 167–179). EMT emphasizes the potential of technology and economic trends in curing environmental ills, as opposed to causing them (see Mol and Sonnenfeld 2000). There are two components underlying this optimism. The first component is the notion that new technology will solve the problems created by old technology.

Critiquing this component, chemical engineer Michael Huesemann and statistician Joyce Huesemann argue that "Techno-optimism is pervasive in our society but hardly justified. [. . .] Despite the serious shortcomings and consequences of past technologies, the public often uncritically accepts new technology, believing that additional and more advanced technology will eventually provide satisfactory solutions" (2011, xxiii). One form of such technological solutions is what they call "counter-technologies, which are technologies specifically developed to oppose and neutralize the negative effects created by other technologies" (2011, 72). Geoengineering schemes proposed to counter climactic effects of excessive carbon emissions exemplify this.

Second is the notion advanced by EMT theorists that production will become less environmentally disruptive as technologies improve (e.g. Mol and Spaargaren 2000; Mol and Sonnenfeld 2000; for an overview and critique, see York and Rosa 2003). This idea that further innovation can create solutions for environmental problems includes the hypothesis of dematerialization, which suggests that through increased efficiency and technological innovation it is possible to decouple "economic growth from the use of energy and materials and from waste flows into the environment, reducing the environmental impact of each additional monetary increment of GDP" (Foster 2002, 22). Marking a departure from his previous work (Thorpe and Jacobson 2013), this perspective is adopted by Giddens in his climate change study, in which he asserts that "greater energy efficiency *ipso facto* reduces emissions" (Giddens 2009, 107).

Foster, however, counters this assumption arguing that dematerialization is a "dangerous myth" (2002, 24). He points to empirical trends of increased consumption of environmental resources as well as increased waste outflows within the most advanced economies as evidence against the notion of dematerialization. Furthermore, to the extent that increased efficiency has been achieved, it "has been invariably accompanied through the history of industrial capitalism by expansion in the scale of the economy [. . .] and hence widening environmental degradation" (Foster 2002, 23). This critique is consistent with the Jevons paradox, based on the finding that technological efficiency improvements that reduce marginal consumption of fuel per unit output lead to overall fuel use *increasing* rather than decreasing (Jevons 1866, 122–137), as well as Richard York and Eugene Rosa's challenge to EMT proponents to demonstrate whether the "pace of increase in efficiency exceeds the pace of increase in overall production" (York and Rosa 2003, 273).

Foster contends that technology cannot solve the environmental problems we face largely because within the capitalist system, "technological change is subordinated to market imperatives" (2002, 38). For example, he points to vested interests standing in the way of increased efficiency such as in the automotive (and corresponding petroleum) industry. From Foster's perspective, integrating environmental issues into the current economic system is a stopgap at best and more likely the source of new problems that arise from commoditizing the environment and subjecting it to the rules of the market (2002, 38–39). The commoditization of nature, according to Foster, necessarily leads to its overexploitation (2002, 39).

For Foster, the capitalist system is inherently incompatible with environmental sustainability (2002, 9–12). Part of this incompatibility, according to Foster, stems from a discordance of temporality. While capitalist investment must be realized within a definite and calculable period of time, issues of biospheric sustainability cannot be incorporated into a short-term temporal perspective (Foster 2002, 10–11; cf. Adam 1998). Furthermore, Foster illustrates the long-standing conflict between powerful economic interests and the environment in which even modest regulations, for instance of GHGs, have been adamantly resisted (2002, 13–22). Since carbon emissions are an intrinsic factor of our energy economy, drastically reducing these emissions challenges the core premise of a capitalist growth economy, at least one built upon fossil fuel energy sources.

Along these lines, Blühdorn and Welsh suggest that the "reassuring belief in the compatibility and interdependence of democratic consumer capitalism and ecological sustainability has become hegemonic," stifling other viewpoints, especially those that are inconvenient (2007, 186). They argue that we have entered a "post-ecologist era," which coincides with a general trend of technological and managerial optimism that coexists with "a fixation on economic growth" and "the normalisation of environmental crisis" (Blühdorn and Welsh 2007, 187). Not only do the "key principles of consumer capitalism, i.e., infinite economic growth and wealth accumulation, which ecologists have always branded as fundamentally unsustainable, remain fully in place," but capitalism has been rebranded as a solution to the ecological crisis (Blühdorn and Welsh 2007, 187). The politics of unsustainability is not characterized by the "denial of environmental problems"; on the contrary, rhetoric and acknowledgment of environmental crises have intensified in recent years (Blühdorn and Welsh 2007, 187). According to Blühdorn and Welsh:

> The politics of unsustainability is unfolding amidst the simultaneity of, on the one hand, a general acceptance that the achievement of sustainability requires radical change in the most basic principles of late-modern societies and, on the other hand, an equally general consensus about the non-negotiability of democratic consumer capitalism—irrespective of mounting evidence of its unsustainability. (Blühdorn and Welsh 2007, 187)

This simultaneous coexistence of paradoxical notions requires either actual or dramaturgical reconciliation.

Given this fundamental contradiction, Blühdorn finds that in the post-ecologist era, the long-standing understanding of "symbolic politics" is insufficient to explain the dynamics of eco-politics (2007, 253). Rather he argues that a new "discourse of seriousness [. . .] adds an *additional layer* of performance" (Blühdorn 2007, 253, emphasis in original). He offers the concept of *simulative politics* as an alternative to the more simplistic idea of symbolic politics to describe modern eco-politics. He finds that "despite their vociferous critique of merely symbolic politics and their declaratory resolve to take effective action, late-modern societies have neither the will nor the ability to get serious" (Blühdorn 2007, 253). Hence the performance of symbolic politics through *simulative politics* is the new mode of trying to "sustain what is known to be unsustainable" (Blühdorn 2007, 272).

The turn from ecologism to post-ecologism is explained as part of a larger shift within politics and society. Blühdorn notes that in "late-modern consumer societies" the trend of materialism has resurfaced in conjunction with both a transversal of *risk society* to *opportunity society* and a transition toward democratic disillusionment following an era of social movements undertaking democratic revolution (2007, 260–261). This convergence of factors has contributed to the emergence of post-ecologism in which the seriousness of ecological threats is recognized, but neither politicians nor constituents can imagine a viable alternative to the unsustainable system. According to Blühdorn, despite political recognition of present environmental threats, even calls for real action are a masquerade since there is no real intent to change the unsustainable system. Rather, a form of simulative politics is played out with the effect, and Blühdorn suggests the tacit intent, of sustaining the unsustainable a little longer.

Extrapolating from Blühdorn, it is posited here that geoengineering constitutes a new strategy toward trying to "sustain what is known to be unsustainable" with the increasingly mainstreamed discussion of geoengineering within the political arena exemplifying "simulative politics" (Blühdorn 2007). With mounting sense of urgency around the environmental threat of climate change, it can no longer be ignored politically, although society has largely demonstrated a collective lack of will "to get serious" (Blühdorn 2007, 253) in terms of making fundamental changes to the carbon-intensive economic system. In the meantime, the prospect of geoengineering also allows for a new manifestation of the political strategy of delaying action pending more research. Moreover, despite protestations by its scientific advocates that it is meant to complement rather than replace emission reductions, geoengineering proposals have the potential to legitimize the de-prioritization of emission abatement, thereby facilitating the continuance of the carbon-intensive energy economy, while fostering a new market for technological research and development. As such, geoengineering is harmonious with the dominant paradigm of continual economic growth (cf. Gunderson, Petersen and Stuart 2018).

GEOENGINEERING AND THE CHALLENGE
OF REFLEXIVE MODERNIZATION

The study of geoengineering provides a contemporary lens through which to examine how society engages with global risk. The extent of risks involved in both climate change and geoengineering as a potential response to it cannot be overstated. Both concern matters fundamental to humanity's ontological security. As argued in reference to climate change:

> Risks of climate change need to be understood sociologically in relation to the radical ontological insecurity that arises from the way in which the existential contradiction [of the human relationship with nature] has returned in a new form. Since climate change is not only a problem of risk, but also poses an existential dilemma, it cannot be merely managed at a technical and pragmatic policy level. The reflexive ethical orientation of life politics is essential if society is to cope with the challenge of climate change. (Thorpe and Jacobson 2013, 101)

These issues become all the more pertinent when applied to geoengineering, which involves new risks of global magnitude. While climate change exemplifies an unintended, although critical, side effect of economic activity, the prospect of geoengineering the planet takes modern risks to a new level. Unlike previous environmental effects that have been mostly externalities to economic activities, geoengineering directly and *intentionally* aims to reorient and manipulate environmental processes.

Geoengineering is unique in that the global risks it would bring with its execution would not be the by-product of other activities but directly related to its implementation. An acceptance of its inherent risks would be required as part of the process. However, like other risks of modernity, geoengineering schemes, despite the detailed research that would inevitably precede them, would be characterized not just by known risks but also "unknown unknowns." According to Beck,

> What differentiates the old nation-state security agenda of the first modernity from the new postnational security agenda of the second modernity is [. . .] the regime of non-knowing, or even worse, not just of known, but above all of unknown non-knowing—of "unknown unknowns." (Beck 2009, 40)

Despite the common notion that the modern era is a sort of "knowledge society," Beck contends that:

> Talk of the "knowledge society" is a euphemism of the first modernity. World risk society is a *non*-knowledge society in a very precise sense (emphasis in original). In contrast to the premodern era, it cannot be overcome by more and better knowledge, more and better science; rather precisely the opposite holds: it is the *product* of more and better science. Non-knowledge rules in the world risk society. Hence, living in the milieu of manufactured non-knowing means seeking unknown answers to questions that nobody can clearly formulate. (Beck 2009, 115)

Intervening in complex planetary systems would necessarily involve such "unknown unknowns."

Contributing a layer of nuance to the discussion of modernity, which is less suggestive of a linear process, Ian Welsh introduces the term "peak modernity" to refer to the mid-century period during which "there was substantive symmetry between the ambitions and aspirations of both political and scientific elites" who "were united behind visions of the planned transformation of society by rational, scientific means" (Welsh 2000, 18). While building upon the theories of Giddens and Beck, Welsh challenges their theories' subordination of "human intervention in major risk domains to science" such that "the importance of the associated cultural practices, social relations and values are effectively sidelined" (Welsh 2000, 24–25). Based on his study of nuclear technology, Welsh finds that "science and technology arise through the efforts of a scientific social movement" and social response, including "ambivalence and more committed public opposition is based in social, cultural, and moral attributes as well as scientific and technical ones" (2000, 31).

How society responds to climate change is and will continue to be emblematic of its relationship to global modern risks and to the idea of "reflexivity" characteristic of "high modernity." The shift seen in the work of Giddens is exemplary of a larger trend in climate politics as it:

> represents a retreat from the reflexivity of high modernity that Giddens theorized in his earlier work. Rather, the approach to climate change in [*The Politics of Climate Change*] reflects the characteristic orientation of simple modernization (Giddens 1994: 5, 42, 80–87): an instrumental approach to nature, faith in technological progress and abstract systems of expertise, and the exclusion of ambivalence and uncertainty. (Thorpe and Jacobson 2013, 100)

Part of this retreat from reflexivity seen in Giddens's scholarship includes his shift away from "life politics" (Thorpe and Jacobson 2013). Giddens's original concept of life politics included "lifestyle decisions that *limit, or actively go against, maximizing economic returns*" (Giddens 1994, 102, emphases in original). While formal politics have moved away from the principle of sacrifice for the sake of a collective purpose, social movements arguably remain the vanguard of the types of values or "life politics" relevant to addressing climate change through social behavior. Environmental social movements have been pivotal in bringing, and maintaining, attention to climate change. Yet, the representative organizations and alliances have largely been left out of official channels of climate politics. While climate activists have generally been kept outside the walls of the various United Nations climate summits, the politics on the inside have been characterized largely by nationalistic maneuvering as opposed to realization of a cosmopolitan moment, simulative politics, and an instrumental approach toward nature indicative of a "retreat from the reflexivity of high modernity" playing out in climate politics (Harris 2013; Beck 2009; Thorpe and Jacobson 2013, 100).

To the extent that post-ecologism characterizes the social relationship with nature, it exhibits a retreat from the reflexivity of high modernity toward a resurgence of an instrumental and dominating approach of first modernity. Geoengineering reflects this instrumental approach premised on the possibility of controlling nature and a culture of technological exuberance in which "there is a remarkable confidence that science and technology will solve the major problems facing humanity, including those created in the first place by technologies. Environmental counter-technologies presumably will solve the problems created by polluting technologies" (Huesemann and Huesemann 2011, 144). However, technological fixes create new risks on top of the contemporary risks that emerged as by-products of modernity, technology, and progress (Beck 2009; Giddens 1991).

THE SCIENTIFIC FRAMING OF
TECHNOLOGICAL RISKS

The constitutive theory of discourse understands language, its conventions, and uses as not only expressing thoughts and ideas about objects but also affecting, redefining, and constructing such objects and their related contexts (Mehan, Nathanson and Skelly 1990; Foucault 2010). The "competition over meaning" in which "proponents of various positions [. . .] attempt to capture or dominate certain modes of representation" has been called the "politics of representation" (Mehan and Wills 1988, 364). In political deliberation focused on "the question of action" responding to a crisis, which narratives "come to prevail will strongly affect" the resulting strategies, policies and outcomes (Fairclough and Fairclough 2012, 6, 17). In regard to scientific research communities, Sarah Parry observes that "by analysing the discursive regularities and strategies within scientists' accounts we can understand not only how specific argumentation is constituted but also how such argumentation produces and is produced by particular social and cultural values" (2009, 94). Discourse around a topic like geoengineering is particularly salient as the concept is still bound to discourse and imaginaries (Stilgoe 2015; Markusson 2013). Particularly, the framing of emerging or controversial technologies can significantly affect their reception (Luokkanen, Huttunen and Hildén 2014; Bellamy 2013; Bellamy et al. 2012; Selin 2007; Parry 2009; Rubin 2008; Brown and Michael 2003; Brown, Kraft and Martin 2006).

Geoengineering is controversial for a number of reasons, including its novelty and the magnitude of its risk. Indicative of the level of risk encapsulated in geoengineering proposals, David Keith, who is among the most influential proponents of geoengineering, has been quoted as stating: "It is hyperbolic to say this, but no less true: when you start to reflect light away from the planet, you can easily imagine a chain of events that would extinguish life on earth" (quoted in Specter 2012). Few technologies parallel this extent of global risk, one being nuclear weaponry. As will be discussed, discursive practices can promote legitimation of such technologies, making them more palatable and acceptable to the public notwithstanding existential risks.

Discourse conventions were of central importance in managing public acceptance of U.S. nuclear policy during the Cold War, especially the conventions of framing nuclear weapons as being for deterrence and for the containment of Soviet expansionism (Mehan, Nathanson and Skelly 1990, 134). The convention of deterrence was developed in response to recognized mutual vulnerability to nuclear weapons and was used to justify policy of maintaining "a strategic retaliatory capability so horrifying in its destructive force that no aggressor would dare launch a strike in the first place" (Mehan, Nathanson and Skelly 1990, 134). In their study of nuclear discourse, Hugh

Mehan, Charles Nathanson, and James Skelly highlight how crucial these discourse conventions were in facilitating and maintaining broad public acquiescence of nuclear policy. When officials within the Reagan administration breached the deterrence convention by talking "publicly about nuclear weapons as a way to win a nuclear war", they "set in motion a chain of events which deprived them of discursive control over nuclear weapons and made it possible to question the fundamental assumptions of the cold war" (Mehan, Nathanson and Skelly 1990, 133, 135).

The importance of establishing and maintaining public acceptance is also clear in other cases of contested or emerging technologies. This was manifestly evident in the case of genetic modification (GM) of food crops in Europe where some forms of GM technologies were more accepted by the public while others were fiercely contended (Healey 2014; Poortinga and Pidgeon 2007). Over time, "public consent" has been increasingly recognized as an "essential element" in the "successful stabilisation" of technological fields (Healey 2014, 12; cf. Markusson 2013, 17). For example, Peter Healey points to the efforts made by proponents of nanotechnologies to learn from what they "saw as the public relations disaster of GM foods" (Healey 2014, 25).

In the case of nuclear armament, discursive conventions developed and entrenched over years of consistent usage came to be paramount in maintaining public acquiescence. Emerging technologies lack the longevity within public discourse for conventions to have become ossified to the same extent. Certain discursive strategies nevertheless can become recurrent and central to the narrative presentation of emerging fields. For example, Cynthia Selin (2007) identifies ways in which nanotechnology promoters engage with the presentation of temporality and the relationship between present and future as discursive strategies that enhance legitimacy of their technological project. Drawing upon the concept of expectations as well as actor-network theory, Selin's analysis is premised on the notion that technology is "the culmination of competing material and linguistic resources in which the technological artifacts (including their representations) have a role in mutually constituting strategies and in aligning interests and visions of the future" (2007, 199). In the case of nanotechnologies, different representations of the technology's future compete for legitimacy and funding.

Discursive presentation at the science-policy interface is also paramount in regard to biomedical research. Nik Brown and Mike Michael identify the framing of "temporal representations of change and the future" in regard to medical research on xenotransplantation (2003, 8). In particular, they note the "metaphor of the 'breakthrough'" as a "pervasive discursive method for organizing narratives about science" (Brown and Michael 2003, 7). Other research points to ways in which the field of blood stem cell research is discursively refreshed over time to emphasize possibilities and differentiate current research from past failures (Brown, Kraft and Martin 2006). As with nanotechnology, discursive presentation of temporality, through expectations, anticipation, and the rhetoric of promise, serves to promote the legitimacy of emerging biomedical fields.

In other bioscience cases, research agendas are promoted by discursive strategies that emphasize technicality, therapeutic potential, and novel cures. For example, Beatrix Rubin identifies how emphasis on "therapeutic promise" and the potential for "novel cures" in the framing of human embryonic stem cell research is used to garner support and stability for this research field (2008, 13). Also, within the field of stem cell research, Sarah Parry analyzes how framing strategies, particularly scientization through "reframing the embryo question as a technical issue rather than a societal one," served to legitimate and endow with perceived authority the work of some scientists over others (2009, 89). As seen in these cases, discursive strategies can materially affect policy and research funding allocation.

Scholars analyzing the field of geoengineering have likewise identified certain discursive strategies and dominant narratives employed by scientists and proponents. For instance, Tina Sikka finds that geoengineering advocates within conservative think tanks employ four discursive frames "to generate support" for the technology: "a claim to scientific neutrality," "technological determinism," "exceptionalism," and "a focus on market-driven solutions as the only way to deal with the impending social, environmental, political and economic fallout of global warming" (Sikka 2012, 167). Likewise, within the ostensibly apolitical discourse of geoengineering assessments, Rob Bellamy identifies the dominance of framings premised on "insufficient mitigation" and the risk of a "climate emergency," which both "posit a central role for geoengineering in tackling climate change" (2013, 1). Moreover, the tendency of assessments to place geoengineering's consideration in "contextual isolation" from alternative approaches to addressing climate change, along with narrow problem definitions and privileging of certain values and assumptions, "produces a limited range of decision options which seem preferable given those framing effects that are privileged, and could ultimately contribute to the [premature] closing down of governance commitments" (Bellamy et al. 2012, 597; cf. Bellamy 2013; Corner, Parkhill and Pidgeon 2011).

Critiquing the "Experiment Earth" dialogue project, Adam Corner, Karen Parkhill, Nick Pidgeon, and Naomi Vaughan identify the foundational role of discursive practices in framing the presentation of geoengineering to public dialogue group participants. These include focusing on "pros and cons of the various technologies" before broaching broader questions of overall desirability, use of the "climate catastrophe" framing, and imagery and language that normalize the technology, minimize the sense of novelty, and purvey a sense of naturalness (Corner, Parkhill and Pidgeon 2011, 12–15, 25–26; Corner et al. 2013). Such framings influenced public participants in the dialogue, with implications in regard to how the framing of geoengineering can influence public perception at a broader level. As will be discussed in subsequent chapters, these various discursive practices continue in contemporary geoengineering discourse and are present within geoengineering policy reports along with other discursive strategies that contribute to constructing particular notions of legitimacy within the field.

NOTE

1. The standard formula used to discuss environmental impact is IPAT, standing for (Environmental) Impact = Population ×Affluence × Technology.

Section II

SCIENTIFIC DISCOURSE AND THE CONSTRUCTION OF LEGITIMACY

Chapter 2

Science Policy Reports and the Framing of Geoengineering

INTRODUCTION TO THE GEOENGINEERING REPORTS OF SCIENTIFIC ACADEMIES

Proposed geoengineering schemes involve scales of technology that would have profound effects at a planetary level.[1] Due to the scale and risk, geoengineering inherently involves science crossing over into law and policy. This section explores the intersection of science and policy with respect to geoengineering through discourse analysis of key scientific reports that are designed to inform policy debates on the potential strategies of countering climate change through climate engineering. Geoengineering science policy reports represent a manifestation of climate engineering's shift toward mainstream consideration and also serve an important role in setting the discursive tone for discussion of the climate crisis and consideration of geoengineering.

Science policy reports translate scientific knowledge and ideas into language that informs political action. They guide policy in the authoritative and "objective" tone of the "voice of science" (Mukerji 1990). In the case of geoengineering, reports are central sites in constructing conceptualizations of the field of research and its technological possibilities. Beyond simply reporting on the state of the developing field, influential reports affect the trajectory of research and development as well as acceptance and acquiescence of novel technologies, especially due to their preeminent positions in informing policy makers, scientific journalism, and, in turn, public perception. Despite underlying reluctance and ambiguities, geoengineering policy reports are found to contribute to the mainstreaming of geoengineering through a conceptualization and linguistic presentation that normalizes novel technologies while constructing hierarchies of legitimacy in regard to pursuing them.

The most influential policy-oriented scientific reports in terms of setting the tone and parameters around geoengineering were the 2009 UK's Royal Society Report *Geoengineering the Climate: Science Governance and Uncertainty* and the U.S. National Academy of Sciences' two-part report released in 2015 by the NRC Committee on Geoengineering Climate, *Climate Intervention: Carbon Dioxide Removal and Reliable Sequestration* and *Reflecting Sunlight to Cool Earth*. The Royal Society and National Academy of Sciences reports are particularly important because of the scientific esteem they are accorded and the extent to which governments look to these scientific institutions for guiding science-related policy. Imbued with authority (Hilgartner 2000, 3; Stilgoe 2015, 104–108), these reports hold a privileged position within the politics of representation of public political discourse where "power and authority are at stake" (Mehan, Nathanson and Skelly 1990, 136).

The Royal Society, founded in 1660 and chartered by the British monarchy in 1663 (Royal Society 2016), has a long-standing role in informing policy and advising decision-makers on scientific matters. In regard to its 2009 report, Jack Stilgoe notes that publication of the Royal Society report "raises the question of how a topic that had been considered unthinkable in polite scientific company only a few years before became worthy of consideration by the world's oldest science academy" (Stilgoe 2015, 103). The Royal Society report stood out as "the preeminent reference point for debates about geoengineering and its governance" (Stilgoe 2015, 103) from the time of its 2009 publication at least until the emergence of the more extensive NRC report in 2015.

Two hundred years younger, but in the same vein as the Royal Society, the National Academy of Sciences was chartered by the U.S. Congress in 1863 with a mandate to "advise the federal government on scientific and technical matters" (National Research Council 2015b, front matter). The two volumes of the National Academy's *Climate Intervention* report were produced at the request of the U.S. government (National Research Council 2015b, ix). With one volume focused on CDR and the other on Albedo Modification, together they represent a comprehensive report on the state of research in the two primary branches of geoengineering (or *climate intervention* in their terminology). The NASEM subsequently released follow-up reports to the original 2015 NRC report on CDR (*Negative Emissions Technologies and Reliable Sequestration: A Research Agenda*) in 2019 and on solar geoengineering (*Reflecting Sunlight: Recommendations for Solar Geoengineering Research and Research Governance*) in 2021.

The topic of geoengineering has surfaced in IPCC reports to some degree since 1990 and explicitly since 1996 and has received increasing attention in recent reports (see Petersen 2014). The IPCC reports are intended to explain the current state of climate science and provide forecasting based on current knowledge. IPCC has made clear in its many reports the importance of mitigation to minimize the most severe impacts of climate change (IPCC 1990; 1995; 2001; 2007; 2014; 2018; 2021). However, geoengineering—or specific forms of it—have been referenced in a number of IPCC reports (1995; 2001; 2013; 2014; 2018; 2021) with IPCC's lower-warming scenarios since the 2014 report relying on the assumption of CDR technologies. While the primary National Academies and Royal Society geoengineering reports are the focus of this analysis, relevant content from IPCC reports as well as other related reports and publications will be considered for comparison as applicable.

The prestige of these institutions as well as the make-up of individuals on the committees producing the reports imbues them with scientific legitimacy and authority (cf. Hilgartner 2000, 3; Stilgoe 2015, 104–108). The reports synthesize current scientific research and knowledge related to geoengineering while also providing policy recommendations as to its future. Unlike reports undertaken by think tanks, research centers, or other non-governmental collaborations, these reports were commissioned by the governments they were written to inform.[2] As such, they hold a privileged position within public political discourse on geoengineering. The 2009 Royal Society report made an appreciable material impact on the subsequent course of geoengineering funding, research, and debate in the UK (e.g., see Owen 2014, 222–223; Markusson 2013, 9; Stilgoe 2015, 103–124). The 2015 NRC report similarly preluded both a shift to political consideration (as will be discussed in chapter 6) and scientific experimentation (as will be discussed in the next chapter). These seminal reports represent a manifestation of climate engineering's shift toward mainstream consideration and serve an important role in setting the tone for discussion of the climate crisis and consideration of geoengineering.

Previous studies have identified certain discursive practices common in presenting and appraising geoengineering proposals, including within the Royal Society report. Nils Markusson analyzes the Royal Society and other high-profile geoengineering reports as "articulations of the geoengineering imaginary" that "construct framings of geoengineering," with consideration of ambivalence within individual reports and diversity between documents in terms of issues like geoengineering's relationship to mitigation as well as its novelty and feasibility (Markusson 2013, 4). He argues "that ambivalence, together with diversity, is [. . .] indicative of attempts at forging new relationships around the geoengineering imaginary" (Markusson 2013, 4). Richard Owen surveys a selection of geoengineering reports and principles, arguing that "the boundary work of experts (e.g., through their visions and judgements) and learned societies (e.g., through their reports), has attempted to legitimise SRM research as an object of governance, specifying certain normative principles and thresholds" (2014, 217). Stephen Gardiner reviews the Royal Society report in regard to issues of ethics, finding that the report "is predicated on a particular account of the ethical context and rationale for geoengineering" and that its "evaluative assumptions [. . .] make substantive differences to policy" (2011, 164). Jack Stilgoe asserts that, despite Royal Society staff and working group members being "admirably open-minded, the issue became scientised in some important ways through their endorsement" (Stilgoe 2015, 15, 94).

While the Royal Society report has been subjected to such analysis, the more extensive NRC and later NASEM reports have not received comparable attention. So, in particular, this section contributes analysis focused on these important reports, examining how discursive strategies are used to construct and reinforce notions of legitimacy in regard to pursuing geoengineering research. Discursive strategies are taken here as communicative devices promoting certain meanings and courses of action, which are motivated by social and material interests. While the question of intents is not part of this analysis, it is acknowledged that actors are not necessarily intentional or reflective in regard to potential secondary outcomes. For example, discursive strategies may be used by actors to promote support of a research program that advances a technology's development even as they remain ambivalent or reluctant about that technology's deployment. It is found that, despite reluctance and ambivalence within geoengineering reports (see Markusson 2013), the reports advance the legitimization, normalization, and mainstreaming of geoengineering. Moreover, certain discursive techniques are used toward advancing a research agenda that would be necessary for the eventual implementation of climate engineering, despite explicit caveats and underlying reluctance of the authors regarding such implementation.

In reference to discourse and language, the NRC report volumes not only are key documents relevant to this topic but also have the advantage of being explicitly conscientious and intentional about their use of language. They reject the terms "geoengineering" and "climate engineering," saying that not only do these terms lack specificity but "the term 'engineering' implies a more precisely tailored and controllable process than might be the case for these climate interventions" (National Research Council 2015b, 1). They also reject the terms "Solar Radiation Management" (SRM) and "Albedo Enhancement" in favor of the more neutral term "Albedo Modification." This choice of terminology is not explained except to say

the Committee chose to avoid the commonly used term of "solar radiation management" in favor of the more physically descriptive term "albedo modification" to describe a subset of such techniques that seek to enhance the reflectivity of the planet to cool the global temperature. (National Research Council 2015b, x)

These linguistic decisions by the authoring Committee demonstrate a reflexivity about language use that implies word choices are conscientiously made in these reports and thus the discourse is indicative of the authors' meaning and intention.

Through detailed analysis of these seminal reports, this section identifies and examines discursive framing strategies that have set the tone of the geoengineering debate and construct notions of legitimacy and normalcy related to geoengineering. This chapter analyzes the entrenchment of discursive trends in how geoengineering concepts and proposals are presented in science policy reports, which are notably written by scientific societies with the purpose of informing policy makers. The discursive trends identified include elevating particular geoengineering methods through comparative evaluation, emphasizing the narrative of geoengineering as a Plan B, and the normalizing and naturalizing of geoengineering proposals through analogy. The following chapter will specifically consider how geoengineering research approaches and actors are presented. This includes themes like the relative legitimization of actors and approaches, differentiating research from deployment, as well as using public engagement strategies and explicit transparency to gain public trust. The analysis finds that public perception management is an acknowledged consideration in the formulation and construction of scientific geoengineering discourse.

It is found that the NRC geoengineering policy report builds upon discursive strategies present in the Royal Society report. The repetition, deepening and entrenchment of these discursive practices contribute to the legitimization, normalization, and mainstreaming of geoengineering research. The subsequent 2021 NASEM *Reflecting Sunlight* report, which grew out of the NRC 2015 report, continues with many of the precedents established by the earlier reports but, in an interesting twist, takes on a sort of reflexivity of this situation, explicitly deliberating on matters of "framing" and "public perception" as will be discussed.

DEFINING/FRAMING THE PROBLEM
AND MATCHING THE SOLUTION

At a basic level, how the problem is defined affects what solutions are considered and how they are evaluated. Of course, to define a problem, it first must be acknowledged as a problem. While there remains political argumentative discourse that attempts to question the seriousness of anthropogenic climate change, this politically and economically motivated argumentation, commonly known as climate change denial, is relevant to scientific discourse only to the extent that it forces scientists to repeatedly reassert that indeed there is a problem to be solved (cf. Oreskes 2004).

As with other climate reports, reiteration of the basic existence of climate change is recurrent in geoengineering reports. Both NRC volumes, for instance, open with setting the stage of climate change and its human causes:

> The signs of a warming planet are all around us: rising seas, melting ice sheets, record-setting temperatures, with impacts cascading to ecosystems, humans, and our economy. At the root of the problem, anthropogenic greenhouse gas emissions to the atmosphere continue to increase. (National Research Council 2015b, ix; 2015a, ix)

Six years later, in the 2021 *Reflecting Sunlight* report, Committee Chair Chris Field writes in the preface: "Despite overwhelming evidence that the climate crisis is real and pressing, emissions of greenhouse gases continue to increase" (NASEM 2021, xi). Then, the Committee's opening line of the summary chapter reads: "Anthropogenic climate change is creating impacts that are widespread and severe," continuing the theme of explicitly reiterating the basic fact of human-caused climate change (NASEM 2021, 1).

In this way, the problem of anthropogenic climate change is re-established from the beginning before the authors can pivot to the matter of how to respond to the challenge. The persistent strand of denialism, especially in the United States, leaves scientists with little choice but to consistently reiterate the existence of anthropogenic climate change in related scholarship and reports even as this forced rearticulation, in effect, plays into the notion that there is an ongoing debate. The political discourse of denialism, along with its effects of undermining and obstructing science, has been critiqued and analyzed by others (e.g., Oreskes and Conway 2010; Buell 2003; McCright and Dunlap 2010). While not the focus of this chapter, the interaction of climate change denial discourse with political consideration of geoengineering will be discussed in greater depth in chapter 6, as it plays out in an interesting way in U.S. Congressional politics.

Within the scope of climate policy discourse there are a number of potential problem-framing options that could be employed. The four framings listed in table 2.1, while not exhaustive, give an idea of how framing the issue

Table 2.1   Examples of Ideal Type Problem Framing and Correlated Solutions

Problem Framing or Phrasing:	Solutions Include
"Global Warming"	Albedo modification
"Climate Change"	Regional adaptation
Elevated Atmospheric CO_2	Carbon dioxide removal and sequestration
Anthropogenic Emissions	Reducing emissions

can influence the range of solutions considered. For instance, if the problem is defined as global warming, within that narrow scope, then albedo modification is a logical engineering solution to consider. If the goal is to counter temperature rises rapidly and globally, most evaluations agree that albedo modification would likely effectively meet this goal. For example, the Royal Society states that although technological readiness for deployment may take some time, "Atmospheric temperatures, however, would respond quite quickly (within a few years) once they were in place" (Royal Society 2009, 32; citation attributed to Matthews and Caldeira 2007; see also National Research Council 2015b, 31).

Geoengineering reports, however, emphasize that there are substantial risks and unknowns that could have extreme environmental impacts, including on ecosystems, hydrological cycles, and other Earth systems that could adversely affect humans and other species significantly. According to the NRC report: "Deploying albedo modification could produce a generally cooler climate, but would introduce risks of a different type. Compensation by albedo modification is only approximate, and some manifestations of high CO_2 concentrations are not addressed at all" (2015b, 33).

This leads to another recurrent caveat that albedo modification obviously has no effect on carbon concentrations. For example, the NRC report states clearly "Albedo modification techniques mask the effects of greenhouse warming; they do not reduce greenhouse gas concentrations" (National Research Council 2015b, 1; cf. 3, 6, 33, 34, 145). The National Academies 2021 report states: "It is first important to recognize that no SG approach can simply reverse the climate effects of increased atmospheric GHG concentrations" (NASEM 2021, 50).

Specifically, it is often mentioned that albedo modification cannot help with the problem of ocean acidification that accompanies global warming as a result of high carbon levels. For example, the 2014 IPCC report states: "SRM would not prevent the CO_2 effects on ecosystems and ocean acidification that are unrelated to warming" (2014, 102). Similarly, the NRC Committee repeatedly notes that "albedo modification does not address the ocean acidification problem" (National Research Council 2015b, 34; see also 2, 6, 145, 146). The Royal Society states,

> It would be risky to embark on the implementation of any large-scale Solar Radiation Management methods, which may not be sustainable in the long term, and which would do nothing for the ocean acidification problem, without a clear and credible exit strategy. (2009, xi)

The 2021 *Reflecting Sunlight* report similarly states that solar geoengineering "does not address the underlying driver of climate change (increasing GHG concentrations in the atmosphere) or the key impacts of rising atmospheric CO_2 such as ocean acidification" (NASEM 2021, 2).

While a focus on warming can be correlated with albedo modification, on the other end of this range, if we consider the problem to be excessive GHG emissions, then the obvious solutions include reducing emissions through converting energy systems away from fossil fuels and lifestyle changes especially in regard to production and consumption practices. A slight adjustment in framing from emphasizing emission activities to the product of these emissions, as in "Elevated Atmospheric CO_2" on table 2.1, opens space for CDR to supplement or supplant emission reduction as a potential solution. The framing of "Climate Change" leaves room to include more localized responses, namely regional adaptation, contrasting from the globality of albedo modification or CDR which sandwich it on table 2.1. However, the framing of the problem and solutions need not be mutually exclusive and might overlap or may be pursued simultaneously, as will be discussed.

In major climate science and policy reports, including those advancing knowledge on geoengineering, the problem framing tends toward the broader framing. For example, "at the root of the problem, anthropogenic greenhouse gas emissions to the atmosphere continue to increase, a substantial fraction of which diffuse into the ocean, causing ocean acidification and threatening marine ecosystems" (National Research Council 2015b, ix). Elsewhere, the report states, "Approaches that limit or reduce levels of CO_2 in the atmosphere address the major cause of human-induced climate change, whereas albedo modification attempts to counter some effects of high greenhouse gas concentrations without addressing the causes" (National Research Council 2015b, 35; see also 3, 145). The 2021 NASEM report similarly states:

> Current scientific understanding makes clear that the changes in climate are being driven by the rapid rate at which greenhouse gas (GHG) concentrations are increasing in the atmosphere. [. . .] Aggressive action to stabilize and reduce atmospheric GHG concentrations can address this problem directly. (NASEM 2021, 31)

The Royal Society indicates carbon concentrations as "the root cause of climate change and its consequences" (2009, 49; see also ix, x, 9). The Royal Society further states: "Increasing atmospheric concentrations of greenhouse gases (chiefly CO_2, with small contributions from N_2O, CH_4, ground level O_3 and CFCs), are the main human causes of warming of the physical climate system" (2009, 9).

However, despite this emissions-focused framing of the problem, the geoengineering reports tend to shift the discourse when turning from explaining the problem to the realm of potential solutions. First and foremost, all the reports emphasize that mitigating emissions is paramount. Ultimately, how-ever, the geoengineering reports recommend pursuing CDR and advancing research on SRM to make that option available and to aid informed decision-making "should it ever be deemed desirable" (National Research Council 2015b, 150) or "necessary" (Royal Society 2009, ix, x, xii, 36, 47, 52, 57, 60).

The emerging narrative that becomes explicit and embedded throughout the NRC report as well as the subsequent NASEM reports is the need for a "portfolio" of responses (National Research Council 2015b; NASEM 2021; NASEM 2019). The 2021 NASEM *Reflecting Sunlight* report recommends that "the United States should implement a robust portfolio of climate miti-gation and adaptation" and that additionally, "given the urgency of climate change concerns and the need for a full understanding of possible response options, the U.S. federal government should establish—in coordination with other countries—a transdisciplinary, solar geoengineering research program" (NASEM 2021, 8). Within this report, the majority of references to the "port-folio" of responses similarly focus on the importance of a multi-pronged approach in which solar geoengineering research does not distract from miti-gation and adaptation. However, upon directly engaging the topic of moral hazard and mitigation deterrence, the Committee notably states:

An expanded research program can be expected to have greater social acceptability if it is embedded within a portfolio of climate policies and research investments that include a firm policy commitment to decarbonization. In the absence of such a commitment, expanded funding for research risks exacerbating moral hazard concerns and reducing societal acceptability of both research and prospective deployment. (NASEM 2021, 121)

This phrasing, emphasizing the importance of the "portfolio" approach in promoting social acceptability and hence increasing the potential for "expanded funding," problematizes whether advocacy for the "portfolio" approach is entirely based upon scientific and ethical optimization or public perception optimization with it being more palatable to the public. However, this particular angle on the importance of the "portfolio" stands out as distinct from the other arguments for pursuing a portfolio approach throughout the report, which implies face value rationales for the importance of mitigation irrespective of solar geoengineering research.

In the NRC's earlier framing of geoengineering options to be considered within the "portfolio" of climate solutions, the Committee says: "CDR methods have more affinity with solutions aimed at reducing net anthropogenic CO_2 emissions [. . .] whereas albedo modification approaches aim to provide symptomatic relief from only some of the consequences of high greenhouse gas concentrations" (National Research Council 2015b, 18). The medical metaphor of "symptomatic relief" aids in bridging the divide from the clearly laid out problem (emissions) to a serious consideration and advancement of a solution (albedo modification) that does not address this problem, but rather addresses one of multiple "symptoms." The medical metaphor provides a sense of legitimacy by drawing upon a well-respected scientific discipline in which treating and alleviating symptoms is understood among both practitioners and the public as an acceptable and desirable standard of care even when such treatment is not curative of underlying causes.

In regard to CDR technologies, the closer fit between problem and solution framing is emphasized by the Royal Society report when it states that: "All of the CDR methods have the dual benefit that they address the direct cause of climate change and also reduce direct consequences of high CO_2 levels including surface ocean acidification" (2009, 21). The Royal Society report defines the problem in terms of elevated or increasing "atmospheric concentrations" of CO_2 and other GHGs (2009, 9, 24, 31, 45, 49) as opposed to "emissions" per se, although emissions, being the primary source of those elevated concentrations, are indicated as a pivotal concern (e.g., Royal Society 2009, v, ix, x, 9, 10, 44, 45, 56, 57). In short, the last two framings listed on table 2.1, elevated atmospheric carbon and anthropogenic emissions, are used to present the problem being discussed in these reports, yet albedo modification (corresponding to the more narrow framing focused on increased global temperature) remains a major part of the discussion of potential solutions being considered.

Various discursive conventions are used to bridge the resulting gap between problem and solution as well as alleviate concerns about the dangers of geoengineering. One convention is the "portfolio of responses" concept consistent throughout the body of scientific policy reports, which allows for the simultaneous advocacy for geoengineering research but while emphasizing that it should not overshadow mitigation and adaptation efforts. For example, the NASEM *Reflecting Sunlight* report asserts clearly that "the starting position of the committee is that SG is not a substitute for mitigation, nor does it lessen the urgency for pursuing mitigation actions" (NASEM 2021, 25).

In addition, as apparent through a close analysis of science policy literature, further discursive conventions have taken root in geoengineering discourse that serve the function of bridging the gap between the recognized problem of carbon pollution and proposed geoengineering solutions as well as ameliorating concerns regarding the latter. These include normalizing the novel through use of metaphor, framing albedo modification as a Plan B or emergency option, constructing a hierarchy of legitimacy in its pursuit, and differentiating research from deployment.

PLAN B ALBEDO MODIFICATION

The recurrent narrative of geoengineering as a possible "emergency solution [. . .] worth researching" in case society is faced with a "climate catastrophe" has been noted by earlier studies as setting up a premise for geoengineering, which effectively limits the scope of deliberation (Corner, Parkhill and Pidgeon 2011, 13). Corner et al. argue:

> Presenting geoengineering as a possible response to a climatic emergency is problematic, especially if linked to the need to conduct research at an early stage, as it provides a very strong framing of necessity, which could artificially enhance the acceptability of conducting research into these technologies. (Corner et al. 2013, 945)

Rob Bellamy argues that the dominant framings within geoengineering discourse, which emphasize "insufficient mitigation" and the risk of a "climate emergency," tend to "posit a central role for geoengineering in tackling climate change" and place its consideration in isolation from alternative approaches of addressing climate change (Bellamy 2013, 1; cf. Bellamy et al. 2012). While scientific experts, including those advocating albedo modification pursuit, acknowledge that it cannot stand alone as a replacement for emissions reductions, the "Plan B" narrative that solar geoengineering may be necessary if mitigation ("Plan A") is insufficient has remained a central trope of geoengineering discourse (cf. Gunderson, Stuart and Petersen 2018). This dominant narrative, which emerged in earlier geoengineering discourse, has remained central in the seminal U.S. and UK science policy reports.

Despite the risks, uncertainties and limitations involved with it, according to the NRC Committee, "There are a number of hypothetical but plausible scenarios in which deployment of albedo modification might be considered" (2015b, 32). Therefore, advancing the research around albedo modification is advocated to make the option available. In the words of the Royal Society: "Because Solar Radiation Management techniques offer the only option for limiting or reducing global temperatures rapidly they should also be the subject of further scientific investigation to improve knowledge in the event that such interventions become urgent and necessary" (Royal Society 2009, 18). Similarly, the NRC report on albedo modification argues for its technological advancement in case of a "climate emergency" in which case "society would face very tough choices regarding whether and how to deploy albedo modification until such time as mitigation, carbon dioxide removal, and adaptation actions could significantly reduce the impacts of climate change" (National Research Council 2015b, 8).

The risk exists that by investing in Plan B, it could become a self-fulfilling prophecy. This is often attributed to the concept of moral hazard, the "risk that research on albedo modification could distract from efforts to mitigate greenhouse gas emissions" (National Research Council 2015b, 8). However, the NRC report asserts:

> The Committee argues that, as a society, we have reached a point where the severity of the potential risks from climate change appears to outweigh the potential risks from the moral hazard associated with a suitably designed and governed research program. Hence, it is important to understand whether and to what extent albedo modification techniques are viable. (National Research Council 2015b, 8)

Particularly the risk exists that the longer it takes to make progress on other climate solutions, the more important speed of execution becomes. Lower risk options require longer lead time; therefore, over time, the range of solutions contracts and the argument for albedo modification becomes more compelling. As the NRC report indicates:

> Should it ever become important for society to cool Earth rapidly, albedo modification approaches (in particular stratospheric aerosol injection and possibly marine cloud brightening) are the only ways that have been suggested by which humans could potentially cool Earth within years after deployment. (National Research Council 2015b, 31)

This theme that albedo modification technology should be developed in case it "should [. . .] ever become important" or "necessary" is a recurrent motif in the justification of furthering its research (National Research Council 2015b, 31). The Royal Society similarly asserts: "Because Solar Radiation Management techniques offer the only option for limiting or reducing global temperatures rapidly they should also be the subject of further scientific investigation to improve knowledge in the event that such interventions become urgent and necessary" (Royal Society 2009, x). Such research is intended to inform decisions of people and their governments to make informed choices regarding potential deployment and improve the quality of deployment techniques if implemented. For instance, the authors of the NRC report assert:

> If future decision makers reach a point that they are contemplating adopting albedo modification, or assessing such an adoption by others, they will need to assess a wide range of factors, both technical and social, to compare the potential benefits and risks of an albedo modification deployment. (National Research Council 2015b, 9, 152)

Furthermore, they state:

If society ultimately decides to intervene in Earth's climate, the Committee most strongly recommends any such actions be informed by a far more substantive body of scientific research—encompassing climate science and economic, political, ethical, and other dimensions—than is available at present. (National Research Council 2015b, 155)

The 2021 report expands upon and deepens the idea that "society" should be actively engaged in the decisions as to whether to advance and possibly deploy solar geoengineering. The authoring committee references "possible 'climate emergency'" in a list of "Rationales for considering SG" (NASEM 2021, 120, cf. 115). However, this newer report distinguishes itself in regard to the "climate emergency" framing by reflectively and explicitly acknowledging it as a form of framing (NASEM 2021, 83, 124). At the same time, though, it follows the same sequencing of the previous reports, transitioning from explaining that solar geoengineering does not address the root causes of the problem to arguing that "given the urgency of climate change" and "given the enormous risks that climate change poses now and in the future," "the U.S. federal government should establish" a "solar geoengineering research program" (NASEM 2021, 8, 31, 145).

While albedo modification research has been promoted in part based on the vague notion of future climate emergency, there is no clear line or definition of what such emergency looks like. In the time since the Royal Society and NRC reports have been written, climate disasters have become increasingly commonplace and the causality increasingly acknowledged. For instance, these words are being written while surrounded by the smoke of California wildfires, something that has become so frequent as to be normalized and *naturalized* with the nomenclature of "wildfire season" in recent years. The "wildfire season" continues to increase in duration and intensity in places like the Western United States, Australia, and even Siberia, while the Amazon seems to burn without seasonal boundaries. In other regions, hurricane season has lengthened, intensified, and expanded in geographical scope. Drought and heat waves dry out some regions (as seen in the record temperatures in the Western United States and Canada in 2021) while unprecedented flooding subsumes others (as seen in Western Germany as well as China's Henan and Hubei Provinces during the summer of 2021), and so on and so forth (e.g., Watts 2021; Victor 2021). It seems that there is no longer a question of whether there will be a climate emergency but rather where the threshold is that would trigger the decision for drastic action.

Scientists emphasize that plausible scenarios in which albedo modification may be deployed include short-term execution. The Royal Society asserts that "Solar Radiation Management methods may provide a potentially useful short-term backup to mitigation in case rapid reductions in global temperatures are needed" (Royal Society 2009, 59). According to the NRC Committee:

> There are a number of hypothetical but plausible scenarios in which deployment of albedo modification might be considered. One scenario is a response to sudden and severe climate change, which is sometimes referred to as a "climate emergency." If, for example, global warming resulted in massive crop failures throughout the tropics [. . .], there could be intense pressure to temporarily reduce temperatures to provide additional time for adaptation. In such circumstances, there could be demands for immediate deployment of albedo modification, even in the absence of a rigorous assessment of the implications or an adequate monitoring system. (National Research Council 2015b, 32)

This articulation of a hypothetical scenario reinforces the "backup" concept that albedo modification could be reserved for a "climate emergency" defined as "sudden and severe climate change" that causes direct effects on human society such as adverse effects on agricultural production. Here, again, it is suggested that such deployment would be a temporary measure to provide time for adaptation. Additionally, it is implied that it might be "pressure" from below, possibly from the general population of regions experiencing adverse effects of climate change, that would push for deployment of albedo modification techniques. Moreover, they indicate such a push for emergency albedo modification may occur irrespective of the state of development of the technology, monitoring systems, and risk assessment. This is related to another central theme in geoengineering discourse: the argument that research should be pursued by legitimate actors particularly due to the risk of presumably illegitimate pursuit of albedo modification technology by others, as will be discussed in the next chapter.

EVALUATION BY COMPARISON

Common within geoengineering discourse across genres, there is a recurrent trend of assessing geoengineering schemes and methods through comparative evaluation (cf. Bellamy et al. 2012, 608; Corner, Parkhill and Pidgeon 2011, 12, 25). Such evaluation by comparison redirects discussion from broader questions of whether geoengineering options may be feasible or fundamentally desirable toward ranking and comparing various options, thereby obfuscating the non-action option (cf. Bellamy et al. 2012; Bellamy et al. 2013; Bellamy 2013; Corner, Parkhill and Pidgeon 2011). It occurs both through comparing albedo modification to CDR and also by comparing specific methods within these categories.

Despite their profound differences, the two categories of geoengineering have historically been intertwined in discourse, development, and evaluation. Over time, scientific academies have increasingly distinguished between solar geoengineering and carbon dioxide removal, but this has been an incremental process. In 2009, the Royal Society addressed both SRM and CDR within the single *Geoengineering the Climate* report. In 2015, the NRC made a step toward disentangling the categories by publishing its *Climate Intervention* report in two volumes, one focused on *Reflecting Sunlight* and one on *Carbon Dioxide Removal*, but both sharing core elements and content, including shared a preface, summary, and introduction chapters.[3] By the time of the 2019 and 2021 NASEM reports, CDR and solar geoengineering were distinctly addressed in separate reports. However, even with increasing divergence in their treatment over time, CDR and solar geoengineering share an origins story and historical connection that has affected both of their trajectories in terms of conceptualization and evaluation.

While the NRC and Royal Society reports urge separate consideration of albedo modification and CDR (National Research Council 2015b, 18; Royal Society 2009, ix; Stilgoe 2015, 121), in practice they tend to draw comparisons between the categories, delineating relative advantages and drawbacks. For example, the Royal Society's "blob chart," a visual representation comparing geoengineering proposals from both categories, has been identified as a problematic component of the report (Stilgoe 2015, 115–120). Choosing two particular axes—"affordability" and "effectiveness" from a number of plausible options—to visually compare possibilities from both categories of geoengineering proposals resulted in a diagram that implied a clear frontrunner among proposals, whereas choosing other equally relevant X and Y axes would have resulted in conflicting impressions (Stilgoe 2015, 115–120; Maynard 2009).

Both volumes of the 2015 NRC report use a comparative table, "Overview of general differences between Carbon Dioxide Removal (CDR) proposals and Albedo Modification proposals" (2015b, 3, 145; 2015a, 3). Distinct from the Royal Society's "blob chart" that visually suggests a hierarchy of preferable proposals, this table lays out a comparative pros and cons list for the broad categories of albedo modification and CDR. In this table and elsewhere, the authors explicitly state that CDR will be judged mostly on the basis of cost while albedo modification will be judged based on risk. Costs of albedo modification deployment tend to be dismissed as negligible despite the costs required of monitoring and evaluation as well as albedo modification's risks potentially translating to costly deferred liabilities.

Within each geoengineering category, many specific comparisons made on the bases of costs, timelines, and risks inform the relative assessment of particular methods and justify the elevation of some proposals over others. Both reports elevate stratospheric aerosols as among the "most promising" options within the albedo modification category. The Royal Society uses the phrase "most promising" three times in regard to stratospheric aerosols. For instance:

> Of the Solar Radiation Management methods considered, stratospheric aerosols are currently the most promising because their effects would be more uniformly distributed than for localised Solar Radiation Management methods, they could be much more readily implemented than space-based methods, and would take effect rapidly. (Royal Society 2009, xi)

This notion of "most promising" methods reorients the framing from risk (negative) to gradations of promise or potential (positive).

The NRC report builds upon the previous elevation of particular methods, including stratospheric aerosol. In justifying which albedo modification strategies the report covers, they explain:

> Rather than discuss every potential means of modifying Earth's albedo that has been proposed, this report will focus on the two strategies that have received the most attention and which may most feasibly have a substantial climate impact: stratospheric aerosol injection and marine cloud brightening. (National Research Council 2015b, 37)

In the albedo modification technical evaluation chapter, the Committee reiterates that they focus on these two proposals "because studies suggest they have the potential to produce a significant cooling and/or they have been discussed more widely in the literature" (2015b, 39). This demonstrates how the elevation of certain proposals may be perpetuated over time based on the degree of previous attention paid to them in addition to the privileging of particular assessment criteria (in this case estimated efficacy at cooling).

The 2021 NASEM *Reflecting Sunlight* report homes in on three methods of solar geoengineering, with continued elevation of SAI and MCB, following the precedent of the earlier NRC report. The third method covered in the newer report is Cirrus Cloud Thinning (CCT), but it is made clear that this possible avenue is less developed and more uncertain. As the Committee states, "Relative to SAI and MCB, CCT has received relatively less attention, and there is relatively higher uncertainty" (NASEM 2021, 49). While significant uncertainty remains for all solar geoengineering approaches, the evaluation by comparison between the considered technologies reorients this uncertainty into relative terms. If CCT is characterized as "relatively higher uncertainty," the implication is that MCB and SAI are relatively less uncertain irrespective of uncertainty and unknowns that remain in absolute, as opposed to relative, terms.

Notably, the 2021 report acknowledges the "possible approach" of "a risk-risk framework, [which] sets the objective of evaluating the benefits and risks of a given action in comparison to the benefits and risks of alternative actions, or compared to no action" (NASEM 2021, 24). This framework in which "the risks of not using SG are compared to those of using it" would be a step toward addressing the concern of risk-benefit assessments that negate the no-action option (NASEM 2021, 77). However, while the report acknowledges the risk-risk framework as "one possible approach" which has been "suggested," they indicate reluctance (or ambiguity)[4] toward this and other comparative risk assessment approaches in practical application, asserting that "in practice such assessments will be extremely complex" (NASEM 2021, 24, 77, cf. 118).

In regard to CDR, the NRC report elevates two methods above others based on their relative "potential" despite significant challenges. According to the report:

> It is important to emphasize that both BECCS [biomass energy with carbon capture and sequestration] and DACS [direct air capture and sequestration], which are the CDR approaches that appear to have the greatest potential for carbon dioxide reduction given the current state of knowledge, depend on the availability of geologic reservoirs capable of accepting and reliably storing massive amounts of CO_2. (2015a, 86)

Describing these proposed CDR methods as having the "greatest potential," implicitly ranks them in frontrunner positions even as this status elevation is prelude to discussion of the methods' significant lack of scalable sequestration options, a challenge that would "require a thousand-fold scale-up of the current CCS [carbon capture and storage] activities that take place today" (National Research Council 2015a, 86). Despite its magnitude, this obstacle, like others, is treated as an engineering challenge solvable through further research and technological development.

The authors seem optimistic about the potential for research and development to solve such challenges, as well as other limitations and costs facing CDR proposals. The report optimistically concludes that "CDR is at an early development stage, and further research and development and emerging technologies may greatly lower costs and increase capacity and deployment readiness, and may thus significantly alter the above conclusions" (2015a, 86). This commentary demonstrates how particular techniques are elevated relative to other options, giving the sense of high potentiality of their realization despite serious limitations and challenges, some of which are comparable to the challenges that have obstructed international action on mitigation in the first place.

Underscoring this issue of evaluating methods by comparison, there is a notable recurrence in geoengineering discourse of what can be characterized as a decoy effect: inclusion of more extreme or controversial proposals, which by comparison tend to elevate the proposals being advanced by the implicit suggestion that these are relatively moderate or reasonable. The decoy effect will be further discussed in Section III as it is a core discursive theme in general audience literature on geoengineering. Within science policy reports, the decoy effect is more subtle and nuanced than within popular media while remaining relevant. Lord Rees's forward to the Royal Society report makes note of the range of options often presented:

> Many proposals for geoengineering have already been made—but the subject is bedevilled by much doubt and confusion. Some schemes are manifestly far-fetched; others are more credible, and are being investigated by reputable scientists; some are being promoted over-optimistically. In this report, the Royal Society aims to provide an authoritative and balanced assessment of the main geoengineering options. (Rees 2009, v)

Rees thus hints at the extreme or decoy options by obliquely referencing the "manifestly far-fetched" schemes that provide a comparison point by which "others are more credible" (Rees 2009, v). This serves to both acknowledge what is here called the decoy effect to a certain extent while also perpetuating it through Lord Rees's own reorientation of the range of options from "far-fetched" to "more credible," thus elevating some options prima facie based upon their relative credibility rather than their inherent strengths alone. Consistently, the presentation of a range of options broadened by inclusion of decoy or "far-fetched" option choices makes the favored options seem more reasonable by comparison. Furthermore, the broad range of options provides a sense of homing in on a direction of pursuit.

INTERNAL EVALUATION AND
ITERATIVE KNOWLEDGE CLAIMS

The 2014 IPCC report, among its plethora of caveats on geoengineering, explains the exclusion of some proposals from examination with the statement that "The scarcity of literature on other SRM techniques precludes their assessment" (IPCC 2014, 102). This raises questions as to whether a proliferation of literature in itself increases legitimacy of solar geoengineering as a field, or particular methods within it, and to what extent it ultimately steers further research and even potential deployment. Within geoengineering-focused reports, specific methods become elevated precisely due to their having been better researched to date. Those topics that are the most researched become favored for further research.

This pattern is evident within both volumes of the 2015 NRC report. Reasonably the Committee states that while "Other approaches have been suggested," their report "focuses [. . .] on techniques for which there is sufficient information to make a preliminary assessment" (National Research Council 2015a, 34). Then, when homing in on favored methods, in addition to evaluation by comparison to alternative methods, the extent of attention a method has received by scientists affects its positioning. For instance, in the *Reflecting Sunlight* report, in setting up the favored options for consideration, the Committee states: "Two more realistic strategies (stratospheric aerosol injection and marine cloud brightening) are then discussed in greater detail because studies suggest they have the potential to produce a significant cooling and/or they have been discussed more widely in the literature" (National Research Council 2015b, 39). While the first reason presented would explain the elevated status of these methods within consideration as well as the increased attention, they add the second reason, having "been discussed more widely in the literature," as an equally pertinent reasoning using the "and/or" conjunction. Similarly, later in the technical analysis section: "The Committee's discussion will focus primarily on injection of sulfate aerosols or their precursors into the lower stratosphere. This is the most-studied technique, and is also the one that most closely mimics the way large volcanic eruptions cool the climate" (National Research Council 2015b, 55). The criteria of being the "most-studied" serves as justification for the continued privileging of the method in this report and future consideration. In turn, the 2021 *Reflecting Sunlight* report, like the 2015 report, emphasizes stratospheric aerosol injection ("the most studied and best understood of the SG approaches proposed to date") and MCB (with "decades of research on aerosol and marine cloud interactions") (NASEM 2021, 34). The 2021 report also considers CCT but with the caveat that "CCT is the least well understood of the three methods considered" (NASEM 2021, 34).

In these reports the scientists want to focus on existing knowledge rather than speculating on less studied alternatives. While it is reasonable to provide attention accordingly to those methods most studied, this practice has the effect of perpetuating the privileged or elevated positions of certain proposals, contributing to the sense of their heightened legitimacy. This may translate into a risk of insular lock-in as existing courses of study within the community reinforce the direction of research and potential deployment. This process is perpetuated by the fact that the authors of science policy analyses rely on the research within the community and the members and advisers are in close relationship.

NORMALIZING AND "NATURALIZING" PROPOSED TECHNIQUES

Normalizing the Novelty of Albedo Modification Proposals

Geoengineering is controversial, among other reasons, because of its novelty. This is especially pronounced for albedo modification methods that would involve global experiments intentionally shifting Earth's radiation balance, changing global climate to completely novel conditions. Many proposed CDR approaches are also novel with far-reaching repercussions. However, studies of public perception have found that people are more favorable to geoengineering projects that are perceived to be more "natural" (Corner et al. 2013; Corner, Parkhill and Pidgeon 2011, 20–22, 26). Hence, framing and contextualizing geoengineering methods in ways that minimize the sense of novelty and normalize the concepts through comparison to common phenomena or activities would be expected to facilitate a more favorable reception.

Corner, Parkhill, and Pidgeon argue that the framing of naturalness has such a strong effect on public reception that "there is a need to ensure that technologies are not associated with the positive notion of 'naturalness' by analogy if, in fact, they are highly artificial" (Corner, Parkhill and Pidgeon 2011, 26). Yet, such framing is recurrent and persistent in geoengineering reports, which make frequent comparisons between climate engineering techniques and natural phenomena. CDR is normalized through emphasizing similarity to natural carbon cycles and processes (cf. Corner, Parkhill and Pidgeon 2011, 21). Two primary analogies normalize the concept of albedo modification: first, comparison to the natural phenomenon of volcanic eruptions and, second, analogy to the mundane experience of pollution.

A favored albedo modification proposal, Stratospheric Aerosol Albedo Modification (SAAM or stratospheric aerosol injection, SAI),[5] which would involve spraying a layer of sulfur-based aerosols into the stratosphere, relies heavily on analogy to the effects of volcanic eruptions. Variants of the terms "volcano"/"volcanoes"/"volcanic" are used 102 times in the NRC (2015b) *Reflecting Sunlight* report main text and appendices. Volcanoes are evoked in two primary manners. First, the volcano analogy is a useful scientific tool because major volcanic eruptions are the closest "natural experiment" that can inform the scientific basis and understanding of SAAM's potential effects. As the NRC explains, "Some volcanic eruptions have injected large amounts of sulfur dioxide gas into the stratosphere, and observations of these eruptions and their impact on climate can serve as natural experiments for testing our understanding of albedo modification processes" (2015b, 59). Second, comparison between proposed human-engineered albedo modification and natural volcanic eruptions may function to normalize and abate fears of albedo modification's novelty and risks.

Beyond being useful analogies, the observed effects of volcanic eruptions have directly influenced, even inspired, the conceptualization of SAAM. According to the NRC:

> The observed cooling following large eruptions provided much of the initial stimulus for the idea that albedo modification could help offset effects of warming due to anthropogenic CO_2 increase, and attempts to model the observed effects of volcanic eruptions can provide some insight into the complexity of the processes and some of the unknowns that still need to be addressed. (2015b, 59)

Similarly, the Royal Society states: "Global cooling has been produced in the past by volcanogenic sulphate aerosols, providing direct evidence that these particles would have a cooling influence" (2009, 29). It is understandable, then, that the volcanic analogy has been particularly important in developing SAAM. The observed effects of past volcanoes, particularly the 1991 eruption of Mount Pinatubo, inform the idea of SAAM and are referenced frequently and often described at length. Scientist authors also display eagerness to amass further data from future volcanic eruptions, which "provide an excellent opportunity to test and improve our understanding of relevant physical processes" (National Research Council 2015b, 59).

In the "Technical Analysis of Possible Albedo Modification Techniques" chapter of the NRC report, the effects of volcanic eruptions are discussed at length. The authors report that:

> Very large eruptions—the size of El Chichón (1982) or Pinatubo (1991)—produce a detectable climate response that can be used to test simulations of both aerosol forcing and the consequent response of climate, but even smaller eruptions—the size of the Sarychev eruption (2009)—can provide a useful test of our ability to observe and to simulate stratospheric aerosol processes. (National Research Council 2015b, 60)

Effects on climate, especially temperature, are emphasized since this is the main objective of volcanic-modeled albedo modification methods. The authors also consider several other effects, including those on the ozone layer, precipitation, photosynthesis, and cirrus cloud variations (National Research Council 2015b, 60–62). These effects, on balance, are treated as neutral, positive, or uncertain and in need of further research.

In this key chapter, however, the potentially catastrophic human risks from albedo modification are relegated to a footnote: "Other eruptions, such as Tambora in 1815, caused global climatic anomalies that led to widespread crop failure and famine" (National Research Council 2015b, 60). This risk of "widespread crop failure and famine" is also referenced elsewhere in the report. However, in both instances within the main text where "crop failure and famine" resulting from the Tambora eruption are mentioned, the subsequent sentence emphasizes limitations of the volcanic analogy:

> Large volcanic eruptions are by their nature uncontrolled and short-lived, and have in rare cases led to widespread crop failure and famine (e.g., the Tambora eruption in1815). However, effects of a sustained albedo modification by introduction of aerosol particles may differ substantially from effects of a brief volcanic eruption. (National Research Council 2015b, 6)

And:

> Other eruptions, such as Tambora in 1815, caused global climatic anomalies that led to widespread crop failure and famine. Overall, it is difficult to compare the injection of an aerosol plume from a single volcanic eruption to repeated aerosol injections that result in a more sustained albedo modification. (National Research Council 2015b, 143)

In these cases, following reference of human suffering resulting from volcanic eruptions changing albedo, the limitations of the analogy are emphasized.

In the NRC report, then, volcanic eruptions are treated as a useful analogy for discussing intended or positive outcomes of SAAM, yet the analogy's relevance is downplayed or dismissed in regard to some major risks. The potential for SAAM to mimic volcanic eruptions in achieving the intended outcome of reduced global temperatures is confidently communicated. In contrast, the potential for disastrous human consequences—namely global famine—is consistently linked to a caveat declaring the limitations of the analogy. Hence, the communication of the volcanic analogy is biased toward the potential positives, while downplaying the potential dangers.

The 2021 NASEM report builds upon the foundation of the volcanic analogue with reference to 2015 report: "As reviewed extensively in NRC (2015), evidence from large volcanic eruptions serves as the essential demonstration that it is possible to reduce solar (shortwave) heating of the planet" (NASEM 2021, 38). Upon establishing this foundation, the later report proceeds swiftly to discussing technical considerations. The report notes that climate models cannot capture all the complexity of the processes, but "models are developing rapidly, and the use of Mt. Pinatubo observations is a key constraint used to evaluate their dynamics and microphysics" (NASEM 2021, 41; citing Sukhodolov et al. 2018). However, it is also noted that "While the study of volcanic sulfur injection has been critical for advancing understanding to date, it is an imperfect analogue for deliberate SAI for several reasons" (NASEM 2021, 41). The condition of this "essential demonstration" being "imperfect" is indicative of the persistence of critical uncertainty and unknowns in regard to solar geoengineering.

A second recurrent comparison, that of inadvertent pollution, is also employed to the effect of normalizing the idea of albedo modification. Addressing concerns about deliberately introducing particulates into the atmosphere through the most elevated options of albedo modification, stratospheric aerosol injection, and MCB, parallel is drawn between the pollution resulting from solar geoengineering processes and inadvertent atmospheric pollution. What is unique to albedo modification compared to other forms of large-scale global pollution is its intentionality. This intentionality is a premise for critique, yet society has become increasingly normalized to the everyday experience of pollution.

Discursively, by linking the intentional release of particulates into the atmosphere with the normalized experience of existing air pollution, there is potential for a subtle shift in the perception of albedo modification's risk or novelty. Indeed, this comparison is emphasized toward the purpose of normalizing the pollutant effect of albedo modification methods. In making the case for outdoor experimentation, the NASEM Committee states that

> the type and scope of the controlled emission experiments that would be required [for MCB research] are very small compared to the nature and emissions of many current human activities for recreation, entertainment, conservation, and commerce purposes, and they are far less than those of ongoing military exercises. (NASEM 2021, 214–215)

Such comparison to the emissions caused by socially accepted activities constitutes a discursive strategy toward normalizing the controversial concept of solar geoengineering experimentation.

The discursive normalization of aerosol pollution is particularly prominent in discussion of stratospheric aerosol injection (SAI or SAAM). The NRC section on the technical analysis of SAAM begins as follows:

> Climate intervention using realistic strategies involves atmospheric injection of aerosols or aerosol precursors. Aerosols (solid or liquid particles suspended in the air) of natural and anthropogenic origin are found everywhere in the atmosphere. They affect the planet's energy budget by scattering and absorbing sunlight, and by changing cloud properties. [. . .] Humans have changed the amount

of aerosols in the atmosphere through pollution emissions, and by changing natural aerosol sources through land and water use. (2015b, 54)

Thus, the discussion of SAAM is framed by the premise that aerosols, both "of natural and anthropogenic origin" are ubiquitous. From ubiquity, it is not a far shift to mundaneness. With this framing, the discourse is flipped, from SAAM as a novel global experiment to a discussion of a commonplace topic (aerosols) simply employed in a new manner (changing albedo). The result is an increasing discursive normalization of SAAM and changed dynamics in the discussion of risk and novelty.

Similarly emphasizing the ubiquity of sulfate aerosols, the Royal Society published a report in 2008 entitled "An Overview of Geoengineering of Climate Using Stratospheric Sulphate Aerosols" which states:

> Sulphate aerosols are always found in the stratosphere. Low background concentrations arise due to transport from the troposphere of natural and anthropogenic sulphur-bearing compounds. Occasionally much higher concentrations arise from volcanic eruptions, resulting in a temporary cooling of the Earth system (Robock 2000), which disappears as the aerosol is flushed from the atmosphere. (Rasch et al. 2008, 4009)[6]

This framing of relevant background information begins with the premise of ubiquity, with an implication of mundaneness, while simultaneously normalizing increased sulfate concentrations with the volcanic analogy in conjunction with the ubiquity framing. In discussion of the environmental impacts of stratospheric sulfur injection, the NRC Committee states:

> Introduction of stratospheric aerosols is likely to slightly increase the acidity of the snow and rain reaching the surface. The effect is estimated to be a very small fraction of the acidity increases associated with industrial pollution today. Thus, any important effects might be counteracted by controlling anthropogenic emissions within the troposphere. (2015b, 75)

In this way, environmental impacts of the resultant pollution are minimized by comparing them to environmental harms already occurring because of human activity. Furthermore, it is suggested that the effects of increased pollution intentionally released by albedo modification could be counteracted by reducing unintentional forms of pollution. This is ironic since the impetus for pursuing albedo modification in the first place is society's inability or unwillingness to adequately control inadvertent pollutions to date.

Analogy to a specific form of existing pollution—ship tracks—is used by the NRC and NASEM committees to explain and normalize the effects of another favored SRM technique, MCB (National Research Council 2015b, 87–90; NASEM 2021, 4, 34, 16, 45, 217). The analogy of ship tracks functions similarly in the discursive and conceptual construction of MCB as does the analogy of volcanoes to SAAM. Scientific analysis of ship tracks has occurred for decades (e.g., Twomey 1977; Conover 1966), and, like the volcano comparison, ship tracks serve not just as a useful comparison in providing relevant data translatable to proposed MCB projects, but ship tracks research has influenced the very conceptualization of MCB and continues to serve as a form of observational study or experiment of opportunity (National Research Council 2015b, 87–92; NASEM 2021, 16, 45, 248).

"Naturalizing" CDR

Just as reports employ discursive framing that normalizes albedo modification concepts, similarly, CDR techniques can be framed to emphasize their "naturalness." Earth's atmosphere is constantly cycling gas compounds through respiration, photosynthesis, geologic weathering, and other natural processes. While CDR proposals model such processes, they differ in being human-engineered with intended outcomes that would redefine the state of balance between relevant natural phenomena.

The NRC report recurrently emphasizes the "naturalness" of CDR proposals by closely aligning them with natural processes. For example: "nature already performs 'CDR' by removing the equivalent of more than half of our emissions from the atmosphere each year" (2015a, 23). Here, CDR is in quotation marks, subtly marking that the analogous natural processes are distinct from the CDR of geoengineering, yet with phrasing that minimizes the distinction. Elsewhere, the Committee comments: "This existing uptake and removal of CO_2 from air, natural 'CDR,' already moderates the impacts of human emissions on atmospheric CO_2 levels and global climate" (2015a, 25). In this way, the NRC report forges a discursive alignment between the CDR of geoengineering and the natural carbon cycle.

This conflation of natural cycles with proposed CDR projects within the NRC report marks a continuation of the theme of emphasizing similarity to natural processes, which was established in the Royal Society report. The Royal Society summary, a critical place for framing concepts since policy makers and others may rely heavily or exclusively on it, emphasized natural processes upon which certain proposed geoengineering techniques are premised, or to which they might contribute, using the word "enhancement":

Enhancement of natural weathering processes to remove CO_2 from the atmosphere. [. . .]

The enhancement of oceanic uptake of CO_2, for example by fertilisation of the oceans with naturally scarce nutrients, or by increasing upwelling processes. . . .

Enhancement of marine cloud reflectivity. (Royal Society 2009, x)

In these instances, the term "enhancement" emphasizes the similarity of these geoengineering proposals to natural processes, minimizing the novelty of the particular endeavors.

Likewise emphasizing the similarity of CDR proposals and natural processes, the NRC states:

There are several CDR approaches that seek to amplify the rates of processes that are already occurring as part of the natural carbon cycle. [. . .] Actions that enhance the reduction of these natural emissions or that increase the natural CO_2 removal from air have the potential to lower atmospheric CO_2. These strategies are variously employed in land management practices, such as low-till agriculture, reforestation (the restoration of forest on recently deforested land), and afforestation (the restoration of forest on land that has been deforested for 50 years or more); ocean iron fertilization; and land- and ocean-based accelerated weathering. (2015a, 28)

Framing CDR as attempts to simply "amplify" and "enhance" natural processes, the NRC continues the discursive trend seen in the Royal Society report. In this selection, the Committee further emphasizes the "naturalness" of CDR by disproportionately including details about the least controversial approaches, namely responsible land management, while leaving the more controversial and novel forms worded with vague technical language. Particularly, reforestation and afforestation are detailed as to what they involve, despite the fact that these terms are likely already more intuitive to a non-specialist audience. Contrastingly, ocean iron fertilization and accelerated weathering are listed without explanation, providing the non-specialist reader less information to make an assessment and requiring them rather to rely on the authors' indication of naturalness. While subsequent chapters of the report further explain these topics, upon the point of arguing their naturalness, they remain abstract.

COMPARING REPORTS

There is a common flow to the discourse of geoengineering policy reports. The "ideal type" of this genre could be summarized as flowing something like this: (1) climate change is a pressing challenge; (2) the preferred solution would be mitigation through emissions reductions, but this has not been adequately achieved to date; (3) anyway, at this point immediate cessation of emissions would not prevent some of the risks of climate change due to the latency problem; (4) due to the risks of climate change caused by continued and latent GHGs, and especially because of the possibility of a "climate emergency," geoengineering should be considered in addition to mitigation and adaptation; (5) there is a broad range of geoengineering schemes that have been proposed as potential options to address effects of climate change; (6) these options all incur risks and costs of varying degrees, some of which are extremely problematic while others could have potential despite the risks, costs, and unknowns; (7) among those options, here is an evaluation based on current research that elevates some and critiques others; (8) with the conclusion that more research should be pursued, especially on those methods distinguished as better than alternative options. The 2015 NRC report is representative of this ideal type, with other geoengineering-focused reports including these topics, albeit with some degree of variation in emphasis and focus.

Each one of these segments in the structuration of policy reports is related to discursive themes discussed within this chapter. Items 1–4 are core to the background and premise of the reports. Item 5 represents the presentation and framing of the "range of options" as discussed here. Item 6 offshoots from this "range of options" framing toward elevating and legitimizing some options relative to others through comparison and utilizing the decoy effect framing, while item 7 addresses internal evaluation and the further elevation (or relative dismissal) of proposals based not only on comparison but also on the level of academic interest in the topic to date. Item 8 directs the overall emphasis on advancing research toward a more narrowed focus resulting from the consideration of items 6–7. The outcome is an argument couched in an implicit sense of inevitability to the prospect of geoengineering, elevating certain options, and promoting the pursuit of the necessary research and development to achieve these options, despite abundant caveats regarding the risk and novelty of geoengineering as well as emphasizing that geoengineering cannot substitute for emissions reductions

All the scientific policy reports strongly urge caution regarding the risks, uncertainties, and dangers of climate engineering. The NRC and Royal Society reports are clear that risks, uncertainties, and environmental impacts make the potential deployment of climate engineering, especially albedo modification, a serious and complex decision. However, they recommend advancing research and development of the technologies in case they "should ever be needed," with the argument that research would allow for informed decision-making and that any deployment should be based on the most rigorous scientific assessment possible.

The IPCC discussion of albedo modification displays a more pronounced reluctance toward its pursuit. The 2014 report provides direct statements regarding the risks, asserting: "If it were deployed, SRM would entail numerous uncertainties, side effects, risks and shortcomings" and that "SRM technologies raise questions about costs, risks, governance, and ethical implications of development and deployment" (IPCC 2014, 102). Whereas the other reports considered arrive at the conclusion that research should be pursued despite these risks and challenges due to the potential of rapid cooling unique to albedo modification, the IPCC report makes a notably more reserved and caveat-infused description of its potential: "SRM is untested, and is not included in any of the mitigation scenarios, but, if realisable, could to some degree offset global temperature rise and some of its effects. It could possibly provide rapid cooling in comparison to CO_2 mitigation" (IPCC 2014, 102).

The 2009 Royal Society, 2015 NRC, and 2021 NASEM reports share many similarities of style and substance, but the time gap in publication, as well as differences of structure and scope, makes for some notable differences. All the reports aim to summarize and explain the current state of the field, drawing upon existing research and relevant publications. A number of the same actors participated in writing and advising on these reports, and many of the same primary sources are relied upon. As such, there was not a pronounced change in general recommendations over the course of each six-year gap in publication. The reports argue that nothing can replace mitigation and that there is no silver bullet to climate change among geoengineering proposals. However, they also argue forcefully for the pursuit of geoengineering research. All highlight the risks of geoengineering techniques and particularly caution that albedo modification techniques are not ripe for deployment. However, they also envision scenarios in which geoengineering strategies may be deemed useful. Indicating that geoengineering is not a substitution for mitigation and adaptation, the Royal Society states that geoengineering "should only be considered as part of a wider package of options for addressing climate change" (2009, 58). Similarly, the NRC and NASEM reports make frequent reference to what they call a "portfolio of climate responses," which would include mitigation, adaptation, CDR, and possibly albedo modification techniques (e.g., National Research Council 2015b, ix, 2, 8, 11, 32, 33, 35–36, 144, 146, 154–155; NASEM 2021, 1, 8, 20–27, 121).

Within this "wider package" or "portfolio of responses," there are some differences in the conclusions of these reports separated by time. The NRC report directly advocates for the pursuit of CDR, while The Royal Society report advocated for continued research but not necessarily pursuit. The Royal Society recommendation that geoengineering "should only be considered as part of a wider package of options for addressing climate change" is followed by the statement that "CDR methods should be regarded as preferable to SRM methods as a way to augment continuing mitigation action in the long term" (Royal Society 2009, 58). In this way, The Royal Society argues that CDR techniques are preferable to SRM but does not quite make an argument for their immediate pursuit. In contrast, the NRC Committee argues that CDR technologies are an integral component of the "portfolio of climate responses." They assert:

> Even if CDR technologies never scale up to the point where they could remove a substantial fraction of current carbon emissions at an economically acceptable price, and even if it took many decades to develop even a modest capability, CDR technologies still have an important role to play. (National Research Council 2015a, 87)

By the time of the 2021 NASEM report, CDR is assumed and built into the portfolio of options:

> Meeting the challenge of climate change requires a portfolio of options. The centerpiece of this portfolio should be reducing GHG emissions, removing and reliably sequestering carbon from the atmosphere, and pursuing adaptation to climate change impacts that have already occurred or will occur in the future. (NASEM 2021, 1)

With that, CDR is now firmly entrenched with emissions reductions and adaption as prima facie portfolio options, and only from there does the NASEM report launch into discussion of solar geoengineering research in case the other three are insufficient.

Another notable change between the Royal Society (and earlier reports) and the NRC report (and later reports) is the reconsideration of terminology. As noted, the NRC Committee in 2015 rejected the terms "geoengineering" and "climate engineering" as well as "albedo enhancement" and "solar radiation management," instead using the terms "climate intervention" and "albedo modification." Such changes have influence beyond semantics since the terminology is integral in issue framing, so terms implying beneficial outcomes (as in "enhancement") or indicating unsubstantiated levels of human control (as in "engineering" or "management") can contribute to a potentially misleading framing of these topics. Even the well-intentioned change to "albedo modification" is arguably problematic in unintentionally obscuring the connotation of risk that has become attached to the term "geoengineering." In the 2021 NASEM report, terminology choices shifted again, this time to "solar geoengineering" in place of "albedo modification." Following the precedent to the NRC, the NASEM Committee also explicitly reflects upon this issue of terminology, concluding that terminology matters in terms of affecting public perception and inviting ongoing consideration on the topic of terminology (2021, 21).

As mentioned in the discussion of evaluation by comparison, the 2019 and 2021 NASEM reports are notably distinct and separate from each other, one on NETs and one on solar geoengineering. This manifests the culmination of the trajectory of separating the two types of technologies, from the Royal Society's addressing both in a single volume to the NRC publishing their report in two volumes, to NASEM's two entirely separate reports. Within these reports, the authoring committees not only follow established discursive conventions but also distinguish themselves from previous reports in certain respects.

The 2019 NASEM *Negative Emissions Technologies and Reliable Sequestration* report embodies the deepening assumption over time that CDR, or NETs as referred to in the report, must be part of climate mitigation policy. While solar geoengineering continues to be a topic of debate, with research advocacy strictly decoupled from deployment (as will be discussed in the next chapter), CDR has been firmly established as part of the "portfolio." Granted, CDR has had a shorter journey with fewer obstacles to overcome in its road to legitimacy. The Royal Society in 2009 recommended that "Carbon Dioxide Removal methods that have been demonstrated to be safe, effective, sustainable and affordable should be deployed alongside conventional mitigation methods as soon as they can be made available" (Royal Society 2009, xi). In 2015, the NRC recommended "research and development investment to improve methods of carbon dioxide removal and disposal at scales that would have a global impact" (National Research Council 2015a, 5). With this foundation undergirding it, the 2019 *Negative Emissions* report is written in a relatively instrumental fashion with a presumption of the inevitability of CDR technologies as a central component in future mitigation efforts. Moreover, it continues and deepens the precedent of the earlier reports in terms of the notion that CDR will be evaluated largely on the basis of cost. The authoring committee states:

> Conclusion 1: Negative emissions technologies are best viewed as a component of the mitigation portfolio, rather than a way to decrease atmospheric concentrations of carbon dioxide only after anthropogenic emissions have been eliminated. The central question is "which is least expensive and least disruptive in terms of land and other impacts—an emission reduction or an equivalent amount of negative emission?" (NASEM 2019, 4)

In this way, NETs are indicated to be not only essential in the "portfolio," but notably it is emphasized that NETs could be treated as interchangeable with emissions reductions, with costs and convenience as guiding factors. They conclude that "the least expensive and least disruptive solution involves a broad portfolio of technologies, including those with positive, near-zero, and negative emissions" (NASEM 2019, 4). The instrumental tone of the NASEM *Negative Emissions* report stands in stark contrast to the NASEM *Reflecting Sunlight* report, which exhibits a particularly high level of reflexivity.

The 2021 NASEM *Reflecting Sunlight* report grew out of the 2015 NRC *Reflecting Sunlight* report and, as might be expected, shares with its predecessor many core themes, emphases, and discursive conventions. It notably differs, however, in two ways, one prescribed and expected and the other more of a surprise. First, the 2015 report recommended that "an albedo modification research program be developed" (Recommendation 4) as well as "the initiation of a serious deliberative process to examine [. . .] research governance" (Recommendation 6) (National Research Council 2015b, 8, 152, 154). These tasks of developing a research program and governance protocols are the raison d'être of the 2021 report. Indeed, in actualizing its purpose, the new NASEM report takes the discourse and conceptualization of geoengineering best practices developed in earlier reports to a new level, building a vision of geoengineering research practices and governance protocols, illustrating how the earlier discourse advocating for such research could become operationalized in practice. Since the development and fleshing out of a research program and governance principles and protocols was the intent of the 2021 report, its contribution to the genre in this way is an anticipated development.

There is another shift, however, that is notably unique about the 2021 NASEM *Reflecting Sunlight* report. Overlaid upon the substantive contributions and embedded discursive conventions continued from the earlier reports is a new level of reflexivity, awareness, and emphasis on the social components of geoengineering research and development. Emblematic of this new emphasis (and also highlighting the report ethos that governance is the key to mitigating risk), the Committee states: "Science and technology do not exist in a vacuum; there are risks associated with the use of any technology, but risk can be moderated by governance" (NASEM 2021, 125). Questions of ethics, social justice, public engagement, and diversity become increasingly interwoven into the scientific analysis. Issues like diversity of scientific representation and meaningful social engagement of vulnerable populations become appended to the existing conceptualization of best practices. As will be further discussed in the following chapter, the 2021 report portrays reflexive considerations on the scientific process and deepened conceptions of public engagement. It explicitly addresses issues like how the framing of geoengineering can affect public perceptions and acceptance of the proposed technologies. In this sense, the 2021 *Reflecting Sunlight* report reflexively integrates and engages with critiques of geoengineering discourse while simultaneously furthering the science policy genre's conceptualization of a geoengineering research program.

This chapter has considered the discourse in geoengineering reports by prestigious scientific institutions, including trends and conventions used in framing particular proposals and conceptions. The next chapter will continue examining these reports and other scientific discourse, with a focus on how the research and researchers themselves are presented to the public, how hierarchies of legitimacy are formed, and how a research agenda is effectively promoted.

NOTES

1. This chapter includes material first published as a journal article (Jacobson 2018).

2. The 2021 NASEM *Reflecting Sunlight* report was not specifically federally commissioned but constitutes a follow-up to the 2015 NRC report which was. Committee Chair Chris Field discussed this nuance in the public webinar upon the report's release (National Academies Webinar 2021).

3. The Preface and Summary sections are fully identical while the introduction (chapter 1) is identical until the final pages of the chapter where the Committee differentiates between the two volumes.

4. Such ambiguity is consistent with Nils Markusson's (2013) analysis of the tensions manifested within co-authored scientific reports.

5. SAAM is the term primarily used in the 2015 NRC report; however, elsewhere a more common term for this proposed technology is stratospheric aerosol injection (SAI), which is adopted in the 2021 NASEM report.

6. This was a review specific to proposed geoengineering with stratospheric sulfate aerosols published a year before the release of the Royal Society's "Geoengineering the Climate" report in *The Philosophical Transactions of the Royal Society A*.

Chapter 3

Science Policy Reports and the Construction of Legitimacy

Research, Actors, and Public Engagement

The issue of legitimacy looms large over discussions of geoengineering.[1] Legitimizing the pursuit of technological imaginaries is integral to the trajectory of moving from margins to mainstream consideration. The construction of legitimacy in the pursuit of geoengineering technologies can be seen to manifest in the reports of elite scientific academies, which in themselves assert a baseline of legitimacy in their very consideration of these topics. How they present proposed geoengineering technologies, research agendas, and best practices constructs a particular vision of legitimacy, while also contributing to the legitimization of the field overall.

RELATIVE LEGITIMATION OF ACTORS AND RESEARCH

All of the key geoengineering reports considered ultimately advocate for the advancement of research endeavors (while always reminding their readers that traditional mitigation remains essential irrespective of geoengineering). The 2009 Royal Society report, 2015 NRC report, and 2021 NASEM report all consistently and pointedly advocate for further research on both climate change in general *and* the potential role of climate engineering. The authors unequivocally support research advancing scientific knowledge related to geoengineering despite recognizing risks of lock-in, vested interests, and moral hazard (e.g., National Research Council 2015b, 123, 125, 129; Royal Society 2009, 45). The 2021 NASEM report, in its characteristically reflexive manner, acknowledges path dependence and how early decisions guide later decisions. The report states:

In designing a research program, it is important to take into consideration that research, technology development, and governance are often path dependent. Early decisions about how to structure and govern SG research may create momentum that shapes future research, development, and governance. (NASEM 2021, 6)

The solution offered to ameliorate the risks of path dependence is a vision of best practices and governance of geoengineering research activities.

While advocating for the advancement of research overall, the 2015 NRC report in particular constructs an implicit hierarchy of relative legitimacy within this research field. Certain protocols and practices related to geoengineering research or potential implementation are indicated, or implied, to be endowed with legitimacy, while others are conversely treated as illegitimate. Closely related is the presentation of relative legitimacy among scientists, national actors, or other groups undertaking such actions. In short, some research and researchers are considered more legitimate than others.

On one end of the legitimacy spectrum, the ideal type would be government-sponsored scientific research, especially "multiple benefit research" with implications for "basic climate science" as well as advancing geoengineering knowledge, conducted using "best practices" and maximum public "transparency" (National Research Council 2015b, 8–9, 11, 112–114, 123, 129, 134, 140, 149–150, 152, 154–155, 209). The other extreme is implied to be rogue actors pursuing albedo modification technology and willing to deploy it unilaterally, particularly to benefit their geographic locality, while putting others at risk for environmental harms (National Research Council 2015b, ix-x, 24, 32–34, 73, 123, 152). The mid-spectrum areas would include, for example, private-sector actors that have vested interests yet adhere to international norms (National Research Council 2015b, 125, 139–140). These various categories of geoengineering actors and practices are conceptualized and presented in ways that indicate their relative legitimacy.

Multiple Benefit Research

Perhaps expectedly, because it is particularly controversial, the NRC's discussion of albedo modification research employs several framing techniques with potential to minimize public opposition and garner support for albedo modification research. One is its strong emphasis on "multiple benefit research" that contributes to the advancement of climate engineering "while simultaneously contributing to the understanding of climate change and other basic research topics" (2015b, 113; see also 8–9, 11, 149–150, 152, 155). Explicitly: "The Committee recommends an albedo modification research program be developed and implemented that emphasizes multiple benefit research that also furthers basic understanding of the climate system and its human dimensions" (National Research Council 2015b, 152). This language justifies further exploratory work on geoengineering by framing the research in terms of the relatively uncontroversial notion that the climate system, climate change, and potential solutions all require further research and then positioning geoengineering as part of that research (cf. Stilgoe 2015, 120–121). This discursive framing neutralizes arguments against geoengineering by presenting its advancement in the form of research, particularly intersecting basic science research, rather than the more controversial framing of implementing a technological program. In this language, geoengineering is brought under the realm of science in its most ostensibly neutral manifestation—basic research—presented through the epistemically privileged "voice of science" (Mukerji 1990).

The notion of multiple benefit research is used to support the case for furthering albedo modification research by emphasizing the benefits apart from geoengineering. For example:

Much of the required research on albedo modification overlaps considerably with basic scientific research that is needed to improve understanding of the climate system. Most notably, research on clouds and aerosols has the potential to advance climate research while also contributing to understanding of the effects and unintended impacts of albedo modification approaches. A number of actions can promote such "multiple benefit research"—research that can contribute to a better understanding of the viability of albedo modification techniques and a better understanding of basic climate science. (National Research Council 2015b, 149–150)

Such emphasis on the breadth of its utility advances arguments for pursuing research relevant to albedo modification.

Due to existing and foreseen objections to pursuing albedo modification, building a case in favor of the controversial research is aided by advocating multiple benefit research, characterized as "research that contributes to albedo modification capabilities while simultaneously contributing to the understanding of climate change and other basic research topics assuming albedo modification is never deployed" (National Research Council 2015b, 113). This emphasis on multiple benefit research particularly speaks to the contingent of the climate science and policy community who have opposed consideration of geoengineering on the basis of its potential to distract and detract from climate mitigation efforts. By including advancement of basic science and climate knowledge, the NRC Committee makes the argument that researching and developing technology relevant to albedo modification "is a no-regrets policy that will be valuable even if albedo modification is never deployed" (2015b, 113).

While emphasis on multiple benefit research enhances the argument for albedo modification research, not all albedo modification research and technology can be classified as multiple benefit. Despite emphasizing multiple benefit where relevant, the NRC authors also argue unreservedly for furthering research applicable solely to albedo modification:

> In addition, there is research that is specific to learning about albedo modification techniques (e.g., mechanisms for delivering sulfate aerosol precursors to the stratosphere) that would not fit under this description of multiple benefit, and is therefore unlikely to be supported without a research program focused on climate intervention. The Committee argues that these research topics specific to albedo modification should also be identified and prioritized as part of a larger research effort on albedo modification, and tasked to the relevant federal agencies for possible support within existing or expanded programs. Focusing on basic science related to albedo modification will hopefully minimize fears that resources are being used to support a potential near-term albedo modification deployment plan. (2015b, 150)

While clearly these arguments for advancing geoengineering research are no doubt based upon genuine concern regarding the risks of unabated climate change and the goal of maximizing climate response options, the scientist authors have a fundamental interest in furthering scientific research as well as a professional culture supporting the notion of scientific knowledge as intrinsically valuable (cf. Stilgoe 2015, 16). This in itself could largely explain the strong advocacy of a research agenda encompassing both multiple benefit and geoengineering-specific research. However, as will be discussed further below, the final sentence of the above-quoted passage indicates a conscious effort to manage public perceptions while advancing an albedo modification research agenda. While a research agenda is not tantamount to a deployment agenda, the language demonstrates an intent to advance an albedo modification research program, while garnering public or political support for it through highlighting other potential research applications.

The emphasis on multiple benefit research is one of the areas in which the 2021 NASEM report differs from its predecessor. Perhaps in part due to the success of the earlier report in making its case regarding the importance of geoengineering research as a component of climate research, the later report demonstrates a greater range of freedom in making its case for the importance of a solar geoengineering research program without relying as much upon arguments for multiple benefit research. This does not mean, however, that the later report abandons the notion of multiple benefit research, only that it is less emphasized. The committee states:

> Some of the information most relevant for policy decisions in this space can contribute to increasing our understanding of basic functions of Earth and its atmosphere, ecosystems, oceans, and societies; however, advancing "basic knowledge" was not the primary driver for the current study. (NASEM 2021, 23)

In this way the research agenda is differentiated from 2015 when basic science and multiple benefit research were more clearly elevated in making the case for albedo modification research.

Yet, even while asserting that solar geoengineering research need not piggyback on other research to be considered legitimate, the idea of multiple benefit research receives due respect in the 2021 *Reflecting Sunlight* report:

> The [research] agenda is also intended to address topics that are not already priorities for the broader climate change research enterprise; although at the same time, we acknowledge (and indeed hope) that some research can advance knowledge both for questions specific to SG and for climate change understanding more generally. (NASEM 2021, 23)

In this way, the later report retains references and deference to the value of multiple benefit research while decoupling the advancement of a solar geoengineering research program from dependence upon its overlapping with other forms of research.

Notably, however, the concept of multiple benefit research is employed when arguing for more controversial positions. While advocating for outdoor experimentation for MCB research, the committee relies upon the multiple benefit research framing in making its case. They state:

> For understanding MCB, the highest priority research questions are how aerosols interact with clouds locally (and immediately) and regionally (and over days). These same questions are also issues of great importance for advancing fundamental understanding of climate change, and thus any advances in this understanding would be beneficial on multiple fronts. (NASEM 2021, 211)

So, even while the legitimating discourse of multiple benefit research is less relied upon generally in the 2021 report, it is clearly applied when advocating for solar geoengineering outdoor experimentation, a topic of greater controversy.

The Layout of Legitimacy

Beyond the classification of research as multiple benefit or specific to climate intervention, the portrayed legitimacy of research is connected to certain notions of how and by whom it is conducted. In this regard, the reports employ terms such as "best practices," "governance," and "international coordination and cooperation." The need for best practices is referenced by the NRC without explicitly expounding its meaning. What is clear is *who* is to lead: "The United States should help lead the development of best practices or specific norms that could serve as a model for researchers and funding agencies in other countries and could lower the risks associated with albedo modification research" (National Research Council 2015b, 11; 2015a, 11). In this manner, the purported legitimacy of scientists working on geoengineering was modeled upon existing structures of global power and influence, premised on the assumption that the United States is a globally responsible upholder of "norms" in the international arena.

Launching from the mandate of the 2015 NRC report, the 2021 NASEM report further develops the notion of best practices, although continuing to avoid explicit definition. It is implied in the newer report that best practices of solar geoengineering research include research that is differentiated from deployment (with "exit ramps" built in), "in the interest of advancing the public good," transparent with international coordination and transdisciplinary participation, with meaningful public engagement (including "key stakeholders" from "climate-vulnerable communities"), subject to regular review by a "diverse, inclusive panel of experts and stakeholders," and "subject to robust governance" (NASEM 2021, 8–9, 71, 138, 145, 181, 254). It is also emphasized by the committee that the solar geoengineering research "program should be a minor part of the overall U.S. research program related to responding to climate change" (NASEM 2021, 145). The United States is still positioned to lead, both by example and through efforts of international cooperation. The NASEM Committee, writing particularly to inform U.S. policy makers, states that "governance mechanisms and principles developed domestically can help inform policy makers developing international architectures" (NASEM 2021, 12). However, even as it is indicated that governance and best practices may originate domestically, efforts at multilateralism are still treated as core to best practices.

Multilateralism is consistently presented as essential for legitimate pursuit of albedo modification, contrasted by the illegitimacy of unilateral pursuit. Although the NRC Committee acknowledges several times that "an international forum for cooperation and coordination on any sort of climate intervention discussion and planning is lacking" (2015b, 7, 148), they indicate that international cooperation will be necessary. For instance, the report summary states: "For the outcome to be as successful as possible, any climate intervention research should be robust, open, likely to yield valuable scientific information, and international in nature" (2015b, 12). The 2021 NASEM Committee further fleshes out the benefits of multilateralism, indicating that international cooperation can help protect against risk of developing an "SG program that only benefits a small minority" and perhaps "magnifies risks for a majority and exacerbates already-existing global inequalities and vulnerabilities to climate change" (NASEM 2021, 184). Moreover, they state:

> International research programs provide opportunities to build trust among parties and open channels for cooperation that may eventually translate into channels for international cooperation on governance. Such partnerships provide opportunities for diffusion of best practices (e.g., through codes of conduct) and protocols for environmental and health safety. (NASEM 2021, 184)

Notably the discussion of the potential for international cooperation on solar geoengineering research reflects an optimism that stands in contrast to the pessimism so often attributed to international cooperation in regard to climate mitigation.

In regard to private-sector involvement, especially of for-profit actors, the NRC report discusses both perceived benefits and risks. The authors point to "known benefits" of private-sector involvement in research, including the ability to "spur innovation, attract capital investment, lead to the development of more effective and lower cost technologies at a faster rate, and produce commercial spin-offs" (2015b, 139). To support this position, they draw on the example of space exploration, saying: "the involvement of private industry contributing to space exploration has generally been viewed quite positively" (2015b, 140). This comparison is given without any explanation or discussion of its applicability. Presumably space exploration is a fitting analogy in terms of ambitious technical and scientific undertakings within disciplines related to engineering and physics. However, there is no acknowledgment of the limitations of such an analogy in terms of the risks and contested desirability of geoengineering, let alone commercial spin-offs, as compared to less controversial aeronautics and space exploration.

The NRC Committee also raises a number of concerns about private-sector involvement, with "the greatest concern [being] that an industry with product lines targeted towards albedo modification would create a group with a vested financial interest in deployment" (2015b, 140). The report quotes the influential *Oxford Principles* stating foremost that geoengineering is "to be regulated as a public good" and that:

While the involvement of the private sector in the delivery of a geoengineering technique should not be prohibited, and may indeed be encouraged to ensure that deployment of a suitable technique can be effected in a timely and efficient manner, regulation of such techniques should be undertaken in the public interest by the appropriate bodies at the state and/or international levels. (National Research Council 2015b, 125; quoting House of Commons Science and Technology Committee 2010; Rayner et al. 2013)

In this way, the risks of vested interests steering the climate engineering agenda are acknowledged and functionally dismissed by delegating such concerns to the realm of a vaguely construed governance protocol that would putatively ensure "public interest."

The 2021 NASEM report similarly approaches the role of private-sector involvement in solar geoengineering research. Also seeing benefits on one hand and risks on the other, the committee states:

Governance of SG research will also need to deal with the opportunities and challenges associated with engagement of the private sector. Private-sector involvement in research and development can spur innovation, attract capital investment, and accelerate the development of effective and lower cost technologies. At the same time, however, there are concerns that for-profit efforts may neglect social, economic, and environmental risks, that research transparency will be compromised by data that are not open and accessible, and that some companies may develop financial interests in moving from research to deployment and seeking private ownership of globally relevant technologies. (NASEM 2021, 25)

Later in the report, the committee grapples with the role of philanthropic support for solar geoengineering research, noting that "At present, more than two-thirds of SG funding in the United States is coming from private sources, including from foundations and individuals" (NASEM 2021, 156). On the one hand, they see benefits of private funding that "may be particularly valuable for advancing research and research governance activities that pose a difficult fit for traditional government funding" (NASEM 2021, 156). But they acknowledge that "there are many concerns about private philanthropy funding SG research," noting that "private sector funding lacks the level of accountability to the broader public typically associated with governmental support" and that there are "ethical [concerns] and other implications of potentially having a small number of wealthy individuals and philanthropies setting research and policy agendas, and shaping the overall path forward, for the SG enterprise" (NASEM 2021, 156–157). Consistent with the emphasis throughout the report that governance can mitigate risks, however, the committee remains optimistic that governance, implementation of "code of conduct" recommendations, and "societal pressure" can mitigate risks related to the participation of private interests (NASEM 2021, 157).

Similarly, irrespective of concerns including vested interests in deployment, the NRC report indicates that there may be desirability in incentivizing private-sector participation, stating:

> A substantial acceleration of albedo modification research would likely require additional incentives, such as public subsidies, GHG emission pricing, ownership models, intellectual property rights, and trade and transfer mechanisms for the dissemination of the technologies. [. . .] These incentives will determine not only whether but how the private sector engages with albedo modification. (National Research Council 2015b, 140; citing Bracmort and Lattanzio 2013)

The NRC's discussion of private involvement is closed with the suggestion that "it would be preferable for the public to have substantial discussion as to what outcomes are desirable before determining what incentives to offer" (National Research Council 2015b, 140). Again here, stated risks are functionally dismissed through reference to hypothetical social involvement of public participation to guide policy, even as the terms of deliberation, and the power structures in funding and guiding research, are being set in such a way that poses challenges to this very public involvement.

Risk of Unilateralism

Contrasting with the implied legitimacy of a multilateral research program is the risk of unilateral pursuit of albedo modification. Emphasis of this risk is recurrent, with the NRC using variants of the term "unilateral" in 20 instances throughout the *Reflecting Sunlight* report. It is highlighted that albedo modification, unlike CDR or mitigation, "could be done unilaterally" (2015b, 3, 145; cf. Royal Society 2009, 40). The committee emphasizes that "A single nation, or even a very wealthy individual, could have the physical and economic capability to deploy albedo modification with the intention of unilateral action to address climate change in a geographic region" (National Research Council 2015b, 122; see also 32). The indication is that there are illegitimate actors who would irresponsibly wield albedo modification technology implicitly compared to responsible legitimate actors.

Echoing the Royal Society's argument that a research moratorium would bind legitimate actors but have no effect on illegitimate actors (Royal Society 2009, 37), the NRC report quotes the argument of David Victor et al. (2009):

> A taboo would interfere with much needed scientific research on an option that might be better for humanity and the world's ecosystems than allowing

unchecked climate change or reckless unilateral geoengineering. Formal prohibition is unlikely to stop determined rogues, but a smart and scientifically sanctioned research program could gather data essential to understanding the risks of geoengineering strategies and to establishing responsible criteria for their testing and deployment. (National Research Council 2015b, 123–124; quoting Victor et al. 2009, 75)

Reference to "rogues" is reminiscent of the concept of legitimate and illegitimate holders of nuclear technology (O'Gorman and Hamilton 2011; Chang and Mehan 2008, 459–461). The loaded political term "rogues" implies entities expected to operate outside of and without approval from the core network of powerful nation-states.

The NRC, like the Royal Society before it, does not explicitly identify who constitutes a rogue threat for geoengineering. However, other government documents highlight China, India and Russia in hypothetical scenarios (e.g., Committee on Science and Technology 2010, 9, 21, 100, 127, 317, 318) and the *Foreign Affairs* article, cited by the NRC in the above excerpt, elsewhere references China and India as nations that may need convincing "not to prematurely deploy poorly designed geoengineering schemes" (Victor et al. 2009, 70). Such a portrayal of legitimacy both reflects and reinforces structures of unequal power in international relations. It is implied that while "rogues" would be irresponsible—indeed "reckless"—with the technology, "a smart and scientifically sanctioned research program" could be trusted to be "responsible" (National Research Council 2015b, 123–124). It juxtaposes the political notion of "rogues" with the presumably politically neutral category of "smart" and "sanctioned" science as practiced within those nations Victor et al. propose should be pursuing geoengineering, although without explicating who all potentially fits into each category.

The risk of unilateralism has entered and persisted in geoengineering discourse, with certain individuals driving its meaning. David Victor and M. Granger Morgan have been particularly influential in guiding discourse on this theme. They have integrated the theme of unilateral deployment into related scholarship (e.g., Victor 2011), used it as a premise for advocating for geoengineering advancement (Victor et al. 2009), and presumably contributed their perspectives on this topic while serving as committee member (Morgan) and reviewer (Victor) for the NRC report. M. Granger Morgan also served as a panelist for a Congressional hearing on geoengineering ("Geoengineering III: Domestic and International Research Governance" on March 18, 2010, which will be discussed in chapter 6).

David Victor, an advocate of geoengineering development, has written about geoengineering and the risk of unilateralism and has come to be called upon as an expert on the subject, being cited frequently as author or interviewee when risk of unilateral climate engineering is discussed (see tables 4.2 and 4.3 in chapter 4). Science policy reports synthesize and reformulate existing discourse, particularly academic but also public discourse, on the topic into the unique package of science policy considerations. On the topic of unilateralism, science policy reports cite to David Victor, in effect, further advancing his status as expert on the topic. The Royal Society notably comments that "concern about the possibility of unilateral implementation has already been expressed by several commentators (eg, Victor 2008)" (Royal Society 2009, 38). Notable here is that David Victor is the only example given for the reference to "several commentators" such that his position is highly represented, perhaps disproportionately so, in carrying the debate on the relevance of unilateralism in geoengineering concerns and policy. In the NRC report, Victor is cited in multiple instances on the topic of unilateralism as well as the related argument that the need for the United States to be on the forefront, setting the norms of geoengineering research (National Research Council 2015b, 23, 123, 124, 140). The later NASEM report also cites specifically, and solely, to David Victor (2019) in identifying "Unilateral deployment" among possible scenarios of solar geoengineering deployment "characterized in the research literature and in broader societal discourse" (NASEM 2021, 200). The NRC and NASEM reports both acknowledge David Victor as a reviewer (National Research Council 2015b, xi; NASEM 2021, ix).

The emphasis on potential unilateralism reframes risk, away from the inherent dangers of albedo modification and instead toward the threat of unilateral pursuit of it. This framing bolsters the case for advancing solar geoengineering research by certain actors through implicit contrast with potential bad actors. It is the climate engineering equivalent to the "if guns are outlawed, only outlaws will have guns" argument. The implication is that albedo modification research by legitimate actors must be supported lest we risk illegitimate actors unilaterally pursuing albedo modification with the mainstream political and scientific community powerless to stop it. This reorients the conception of risk in regard to geoengineering from *what* to *who*. Rather than the primary risk being deployment and its adverse consequences (known and unknown), the risk is redefined and reoriented from deployment itself to unilateral deployment.

The NRC Committee argues that due to the risk of unilateral deployment, research should be pursued even if—perhaps especially if—albedo modification is assessed to be undesirable because the research to develop the technology is also necessary to identify its use by unilateral actors. For example, within their policy recommendations, the committee promotes development of a new generation of space-based instruments that "would significantly improve understanding of the effects of clouds and stratospheric aerosols on climate, improve the ability to predict the effects of albedo modification, and provide an ability to detect large-scale albedo modification by rogue actors" (2015b, 152). Thus the argument for advancing research and technological development can be supported by several layers of reasoning: overlap with basic research (multiple benefit research as discussed earlier), the potential benefit of pursuing albedo modification, or conversely the risk of albedo modification and the benefit of detecting its use by others.

There is a curious parallel between the discursive framing of legitimate geoengineering pursuit and the Catholic notion of Just War Doctrine. Five of the core tenets of Just War Doctrine are that a just war can only be waged by (1) a competent and legitimate authority (2) with right intention, (3) as a last resort with (4) a probability of success and (5) anticipated proportionality of benefits outweighing harms. Analogous discourse is used to justify the pursuit of geoengineering. As seen in the preceding analysis of scientific discourse in science policy reports, the pursuit of albedo modification is delineated as legitimate only if pursued by recognized legitimate and multilateral actors with right intentions and best practices. Moreover, legitimate deployment rests on probability of success and anticipation of benefits outweighing harms. The notion of last resort is discursively elevated while in praxis, a very sticky point in light of the inadequacy of international GHG abatement efforts to date. In the politics of war, the thresholds for meeting these just war requirements and specific circumstances' instantiation of them can be debated and deliberated to claim a particular war is just or not. The deliberation itself constitutes a key part of the justification process. Similarly, the discourse developed to identify legitimate pursuit of geoengineering is also the discourse that builds the case justifying such pursuit.

Differentiating Research from Deployment

The discursive differentiation of research and deployment is central to managing perceptions and legitimating geoengineering action. Within the proposed forms of geoengineering, a full research program would necessarily blur lines between research and deployment (Stilgoe 2016, 858–859; Robock et al. 2010; Hulme 2014, 95). However, studies of public attitudes toward geoengineering have shown that people "tend to make a distinction between research and deployment" and be more favorable "to the idea of researching geoengineering, while holding significant reservations about ever deploying it" (Corner et al. 2013, 941). Hence, simultaneously advocating for research (that would be necessary for deployment) while cautioning against deployment increases the potential of advancing research with minimized impediment. The Royal Society contributed toward developing a narrative distinguishing between research and deployment (Owen 2014, 223).

The 2015 NRC report significantly deepened the distinction, and the 2021 NASEM report followed precedent.

The discursive convention of differentiating research from deployment specifically encourages the advancement of technology while simultaneously discounting the need for social acceptance of its eventual use. Especially in regard to solar geoengineering, the reports advocate for further research while discouraging deployment "at this time." Therefore, social acceptance of research *not* deployment is encouraged, even for research that would create the technological basis for deployment. Focusing the question of social acceptance onto the narrower realm of research pushes the question of social acceptance of deployment to a later date when the technology would be more developed. Moreover, due to geoengineering proposals' global scale of intended consequences, the full extent of consequences would be unveiled only through the global experiment of deployment (Stilgoe 2016, 858–859; Robock et al. 2010; Hulme 2014, 95; Owen 2014, 16).

The NRC report consciously differentiates albedo modification research from deployment. The committee explicitly distinguishes its research advocacy from its position on deployment, with the core recommendations on the subject being:

Recommendation 3: Albedo modification at scales sufficient to alter climate should not be deployed at this time. [. . .]

Recommendation 4: The Committee recommends an albedo modification research program be developed and implemented that emphasizes multiple benefit research. (National Research Council 2015b, 7–9)

In these sequential recommendations, the committee distances itself from the topic of deployment while advocating development of a research program that would be the basis of deployment. As discussed, the emphasis on "multiple benefit research" helps to reconcile this juxtaposition.

Throughout the NRC *Reflecting Sunlight* report, this strategy of differentiating research from deployment is central and recurrent, including explicit statements such as "The Committee reiterates that it is opposed to large-scale deployment of albedo modification techniques, but does recommend further research" (2015b, 155). The chapter on research governance opens by stating this official position: "The focus of this chapter is on the issue of governing research, because research is the only albedo modification-related activity that the Committee believes should be considered at this time" (121).

Further reinforcing this discursive decoupling of research and deployment is the introduction of the concept of "large-scale deployment." The committee states: "There are many research opportunities that would allow the scientific community to learn more about the risks and benefits of albedo modification, knowledge which could better inform societal decisions without imposing the risks associated with large-scale deployment" (2015b, 8). Elsewhere the Committee reiterates that "it is opposed to large-scale deployment of albedo modification, but does recommend further research" (2015b, 155). In this way, the differentiation between research and deployment is subtly shifted to the differentiation of research and "*large-scale* deployment (emphasis mine)." While there is "large-scale deployment," there is not corresponding reference to *small-scale deployment*, but rather "small field studies" (2015b, 151), "small-scale field experiments" (9, 152), "small-scale projects that inject materials into the stratosphere" (81), "small-scale controlled emissions studies" (102), "small-scale experimental studies" (127), and "small-scale experiments" (132, 139).

Only once is the term "deployment" paired with the concept of "small-scale." The committee states that recommended "research encompasses a range of activities from the innocuous, such as modeling, to the more invasive, such as controlled small-scale test-deployments for experimentation purposes" (2015b, 121). While this statement in one sense bridges the gap between research and "large-scale deployment," it also emphasizes the differentiation of research and deployment, especially as it immediately follows a reiteration that only research is being advocated. This research, however, includes "small-scale test-deployments," which are treated as "research," not "deployment," despite their simultaneous presentation as both. The framing emphasizes that these "test-deployments" would be "controlled" and for "experimentation purposes," indicating that this type of experiment would be considered within the research category rather than the deployment category. In this way, the committee is able to assert the broad take-away point that "at this time" they advocate for research but not deployment, implicitly defined as "large-scale deployment," while subtly introducing the notion of "small-scale test-deployments" within the category of research despite the nebulous boundaries between this type of research and deployment.

The ambiguous relationship between the concept of deployment and categories of scale can be seen elsewhere when the committee implies that small-scale deployment, though not phrased as such, must pave the way to actual deployment:

> Any albedo modification, if deployed, should start with an intervention of small magnitude [. . .] in order to gain experience with the consequences of a more modest intervention and its impacts on both to the shortwave energy balance and to other aspects of the system before making a decision as to whether the risks involved in scaling to larger values are tolerable. (2015b, 108)

This *if-should* phrasing allows the authors to maintain two contradictory positions: they are not advocating deployment but rather specifying necessary mechanisms and requirements *if* deployment were to occur, while advocating the achievement of those same deployment prerequisites as necessary research independent of the pursuit of deployment. In this way, even while indicating that the forms of research advocated are those that would be necessary in order to achieve deployment, the authors are able to maintain the explicit decoupling of research and deployment, diminishing the sense that pursuing research would necessarily advance deployment.

Promoting relevant research while simultaneously cautioning against its deployment is at face value—as presumably intended—a pragmatic and moderate position. However, another function for the recurrent differentiation of research (promoted) and deployment (not promoted "at this time") is public perception management. As mentioned previously, this intent is overtly revealed within the NRC report wherein explaining the emphasis on multiple benefit research when the committee states that "Focusing on basic science related to albedo modification will hopefully minimize fears that resources are being used to support a potential near-term albedo modification deployment plan" (2015b, 150). This does not say that the focus on basic science would be due to the unlikelihood or undesirability of deployment, but rather that it would "minimize fears" presumably among the public. (Notably, the more condescending term "fears" is used to characterize potential negative response from the public rather than the more neutral term "concerns." The term "fears" is generally indicative of an emotional and often irrational response, while "concerns" would conversely imply some extent of rational consideration of potential negative outcomes.) This statement indicates an effort to promote albedo modification research while distancing it from the obvious corollary of potential deployment in order to minimize public concern, which is, of course, essential in minimizing opposition.

This framing of research versus deployment may reflect lessons learned in the aftermath of the Royal Society report and subsequent proposed experiments, such as the planned (and ultimately canceled) Stratospheric Particle Injection for Climate Engineering (SPICE) technical experiment, which drew considerable public opposition (see Owen 2014, 236; NASEM 2021, 131, 173). This experiment would have involved the seemingly benign spraying of water into the atmosphere to test what dispersion might look like in albedo modification scenarios, but despite the idea that it would be an "uncontroversial" experiment, it was canceled after significant internal deliberation and also an outpouring of opposition from the public, which culminated in a petition from over 50 organizations voicing opposition (Specter 2012; Stilgoe 2015; Schneider and Fuhr 2021, 63). Opposition included concern over vested interests, especially the involvement of scientists holding patents for relevant technology, as well as concern over the moral hazard issues associated with the advancement of geoengineering pursuits to the potential detriment of mitigation efforts (Specter 2012; Lukacs 2012; Cressey 2012; NASEM 2021, 107, 173, 177). Similarly, within the category of CDR, public outcry has directly affected the moratorium on ocean fertilization experiments and caused the cancelation of specific experiments (e.g., Goodell 2010, 154–158; Kintisch 2010, 140–148; Schneider and Fuhr 2021, 63–64). Concern regarding vested interests that would have a stake in the eventual deployment of these technologies, as well as concern that experimentation adds to the likelihood of deployment, has been central to public opposition to field experiments related to both solar geoengineering and ocean fertilization. In the wake of such public opposition, the subsequent NRC report, written collaboratively with scientists interested in advancing research and experimentation, explicitly decoupled research from deployment of geoengineering methods.[2]

While the NRC Committee displays mindfulness of public perception in its decoupling of research and deployment, the earlier Royal Society report treated the distinction in more pragmatic scientific terms. Although also advocating research and not deployment, the corollary relationship was more directly recognized. Research, not deployment, was promoted in large part because research would be needed prior to potential deployment. Regarding CDR, the authors state pragmatically that "significant research is [. . .] required before any of these methods could be deployed at a commercial scale" (Royal Society 2009, 21). The report similarly acknowledges the relationship between albedo modification research and deployment stating:

> None of the principal proposals are yet ready to be put into operation. Further research and development of the individual approaches (including, in some cases, pilot-scale trials) would be needed to assess uncertainties about effectiveness and undesired side effects and to identify any preferred approach. (Royal Society 2009, 34)

Likewise, "none of the methods assessed are yet ready for deployment, and all require significant research including in some cases, pilot scale trials, to establish their potential effectiveness and effects on climatic parameters" (Royal Society 2009, 36). The relationship here between research and deployment is that research is a necessary stage to either become ready for deployment or to make educated decisions about deployment. The Royal Society's framing is distinct from the later NRC report, in which recurrent use of discursive decoupling serves to minimize the corollary thread between research and deployment readiness.

This newer convention of discursively decoupling research from deployment increases potential receptivity to albedo modification research. However, there is a very complicated interconnection between research and deployment.

Mike Hulme similarly points to the problem of drawing "too clean a distinction between researching and deploying the technology" (2014, 68). Hulme argues: "Once one is started on a course of technological development, it becomes increasingly difficult to stop its eventual deployment" (Hulme 2014, 68). In this way, the evolution of scientific discourse distinguishing between research and deployment supports the possibilities of advancing research, which in the end may ironically increase the probability of deployment.

Following the precedent of previous reports, the 2021 NASEM *Reflecting Sunlight* report also proactively endeavors to differentiate research from deployment, notably varying between the type of forceful conceptual differentiation established by the NRC in 2015 and the more practical differentiation seen in the earlier Royal Society report. Recognizing the practical relationship between research and deployment, the committee states,

> Before implementation of any SG strategies would ever be considered, the potential impacts of any given approach must be understood to the fullest extent possible—hence one of the central goals of SG research is to predict how the climate would respond to a hypothetical deployment. (NASEM 2021, 50)

Yet, in this report, which was specifically charged with detailing a vision of a research program and related governance, they reflexively acknowledge the risk of research paving a path to deployment and explicitly state the intention to avoid such path dependence.

In the chapter laying out the goals and approach of their proposed research program, the committee states,

> A central feature of a national SG research program and of U.S. input into international or other programs is that the goal of the program should be clearly and unequivocally to understand the prospects and limitations of SG options and not to drive toward eventual deployment. (NASEM 2021, 140)

They then go on to state:

> A national research program should be designed to explore the full range of issues relevant to possible future deployment. This should include not only issues related to technical feasibility and efficacy but also issues related to indirect effects, social implications, human perceptions, and judgments about equity. If these technologies are ever seriously considered for deployment, the perceived legitimacy of the research program will be as important as the specific findings. Thus, a key challenge is to develop and coordinate a research program that is informing decisions without committing to further development of that

technology or creating research communities that are invested in its ultimate deployment. (NASEM 2021, 140)

Hence, the 2021 Committee makes clear that the proposed research program should inform decisions regarding deployment, while safeguarding against necessarily leading to deployment. In this sense, the proposed research is set up as a necessary but not sufficient condition to the possibility of eventual deployment. Moreover, they assert that the research program is meant to inform decision about deployment and, very explicitly, that the evidence could point "in favor or disfavor of SG deployment" (NASEM 2021, 6). They state that "a principal goal of any research program should be to better characterize and reduce scientific and societal uncertainties concerning the benefits and risks of SG deployment (relative to global warming in the absence of SG)" and that "it is possible that additional research may expand particular uncertainties or reveal new uncertainties," such that research could potentially develop a stronger case against deployment (NASEM 2021, 6).

Yet, the committee also recognizes challenges to differentiating research from deployment:

We recognize that it may be difficult in some cases to draw the line between feasibility-oriented research and deployment-related research. The best protection is likely to be a robust decision-making process for deployment that can minimize any inappropriate influence stemming from potential profits. (NASEM 2021, 144)

One pragmatic practice which the committee recommends is establishing "exit ramps" from lines of solar geoengineering pursuit should research indicate that deployment is undesirable or unproductive. They argue: "This possibility of exit ramps helps address the general problem of research funding for a specific project or a larger program becoming locked into place and renewed year after year even in the absence of meaningful progress" (NASEM 2021, 141). They also argue that funding should encourage "creative thinking while avoiding commitments to further development of a specific technology or to the creation of research communities that are invested in its ultimate deployment" or, in other words, to avoid technology lock-in or path dependence (NASEM 2021, 144). The Committee recommends that funding ramps up over time and that a significant portion of the budget be "dynamic" funding giving flexibility to pursue new paths or shift priorities as research brings new information and recommends that "exit ramps need to be built into funding plans in order to accommodate the possibility that, based on findings from the research, some (or even all) lines of inquiry may at some point be defunded" (NASEM 2021, 254).

More than the previous reports, the 2021 Committee explicitly acknowledges and engages with the "line between technology capability for research and technology capacity for deployment," asserting:

> The proposed research program does not include the goal of supporting technology development that is specifically oriented toward building the capacity for SG deployment. Yet the development of some specific technology capabilities is needed to advance fundamental understanding of particular scientific questions proposed in this research [. . .] or to better understand the technical feasibility challenges of particular approaches. (NASEM 2021, 230)

For example,

> some development of specific capabilities may be necessary simply to enable scientific experiments. For instance, for MCB studies, there will be a need for spray nozzles that can produce a particular range of aerosol size distributions. For SAI, even small-scale tests of solid aerosols will require some dispersal capability. (NASEM 2021, 232)

Yet, displaying the characteristic reflexivity of this committee, they are mindful of the risks of mission creep and the fine line between technological development and deployment capability:

> Efforts to advance these technical capabilities to conduct SAI experiments, and similarly the efforts to advance MCB dispersion nozzle design, would be focused on facilitating the "atmospheric processes" research discussed earlier in this chapter not providing pathways toward deployment. By focusing the technology-related work on these critical research questions, one is likely to reduce the potential for mission creep toward developing deployment capability. (NASEM 2021, 233)

There are several points where the committee references "the slippery slope toward deployment" (in fact, 18 instances in the main text use the term "slippery slope"), showing an acknowledgment of the concern that research pursuit can potentially increase the likelihood of deployment through a sort of path dependence. Several of their more in-depth engagements with this concept are in the context of anticipating public concerns and ways to mitigate these concerns. For example, they state, "To avoid concerns over research representing a 'slippery slope' toward deployment, simultaneously developing the capacity to deploy as part of conducting scientific experiments is not recommended" (NASEM 2021, 232–233). Elsewhere, funding for SG research should not shift the focus from other important global climate change research, and it should recognize the risk of exacerbating concerns about a slippery slope toward deployment. This guideline implies that the near-term budget for SG research should be small relative to the overall investment in global change research. (NASEM 2021, 251)

In these instances, and others, the emphasis on public perceptions, specifically public concerns regarding a pathway to premature deployment, makes it unclear whether the primary reasoning for differentiating the research program from "the potential for mission creep toward developing deployment capability" is based upon the risk of "exacerbating concerns about a slippery slope" or the slippery slope itself in terms of the risk of technological lock-in (NASEM 2021, 233, 251).

TRANSPARENCY, MULTILATERALISM, AND PUBLIC PARTICIPATION

There is broad acknowledgment in Science and Technology Studies (STS) literature that public acceptance and consent beyond the community of experts is essential for the successful advancement of technological fields in democratic societies (e.g., Nowotny, Scott and Gibbons 2001; Wilsdon, Wynne and Stilgoe 2004; Stirling 2008; 2014; Healey 2014). Yet, a changing societal relationship to expertise, including the so-called "crisis in public confidence," has led to "a new institutional body language for science, including consensus conferences, public participation exercises, science shops and, most notably, the language and rhetoric of transparency" (Brown and Michael 2002, 262). Moreover, the process of interacting with the public regarding scientific and technical issues of social importance has changed over the years from the earlier "deficit hypothesis" of "public engagement" centered on scientists explaining concepts to the public with the assumption "that if only people knew more about a technology, they would come to see its benefits as outweighing its risks" thus reducing public opposition to technologies presumed to be based on lack of understanding and the currently favored "process of dialogue between scientists and the public" (Corner, Parkhill and Pidgeon 2011, 7).

Whether genuine two-way dialogue is established or not, science policy reports in regard to climate intervention and its research governance repeatedly emphasize the importance of transparency, multilateralism, and public participation in the process (e.g., National Research Council 2015b; Asilomar Scientific Organizing Committee 2010; Royal Society 2009; NASEM 2021). These terms are problematic in being vague and subjective. It is unclear what criteria would be used for evaluating the success of these objectives. However, the act of repeatedly articulating them as requirements for ensuring a legitimate process toward climate intervention acts to claim that legitimacy through stated intent, while implying that there is some other illegitimate alternative process (presumably the oft-referenced although never reified unilateral deployment threat) from which the proposed research differs.

Reference to the importance of "transparency" in best practices of geoengineering research is a particularly central theme of these reports, and attention to it has increased significantly over time. For example, the terms "transparency" or "transparent" are used 10 times in the 2009 Royal Society report, 22 times in the 2015 NRC *Reflecting Sunlight* report, and 74 times in the 2021 NASEM *Reflecting Sunlight* report. Certainly, transparency is important when dealing with the pursuit of geoengineering. There are two potential functions of emphasizing transparency. The first is the face value function as communicated: transparency so the public can provide checks against the progress of something that affects them. The other is the opposite: using an emphasis on transparency to minimize public concern. Due to explicit and conscious concern regarding ameliorating public concern, there exists the risk that the latter could overshadow the former function (cf. Felt and Fochler 2010; Stirling 2008, 264, 267; Corner, Parkhill and Pidgeon 2011, 24).

No doubt, the discussion of transparency, multilateralism, and public participation is to a large extent to be taken at face value, namely that these elements are important for "best practices" and ensuring motivations and outcomes that maximize utility and minimize harm. For instance, the 2015 NRC report states,

> Given the perceived and real risks associated with some types of albedo modification research, open conversations about the governance of such research, beyond the more general research governance requirements, could encourage civil society engagement in the process of deciding the appropriateness of any research efforts undertaken. (National Research Council 2015b, 10, 153)

Furthermore, the committee indicates that "Ultimately, the goal is to ensure that the benefits of the research are realized to inform civil society decision making, the associated challenges are well understood, and risks are kept small" (National Research Council 2015b, 10, 153). This statement manages public perceptions by framing the concept of geoengineering going forward as a process which would be controlled by civil society. Furthermore, the major issues of uncertainty and unknown risks that are emphasized when actually focusing on the science of geoengineering are minimized in this public engagement framing, which indicates "challenges" *can be* "well understood" and that "risks" *can be* "kept small."

However, the abstract notion of public participation, while aiding in discursive claims to legitimacy, does not necessarily translate into an engaged democratic process (e.g., Stirling 2008; Brown and Michael 2002; Corner, Parkhill and Pidgeon 2011; Felt and Fochler 2010). Rather, discursive emphasis on the concept of public participation can also be utilized in a more subversive sense. "Public participation" carries connotations of democratic engagement of society and the notion of citizen oversight. However, the way the term is used in geoengineering reports at times signals an intent to utilize public participation strategically in order to placate the public in regard to the pursuit of climate engineering endeavors. Similarly, the terms multilateralism and transparency are often employed in a manner relevant to perception management.

Public perception is treated as a challenge to the pursuit of geoengineering. According to the Royal Society:

> Public attitudes towards geoengineering, and public engagement in the development of individual methods proposed, will have a critical bearing on its future. Perception of the risks involved, levels of trust in those undertaking research or implementation, and the transparency of actions, purposes and vested interests, will determine the political feasibility of geoengineering. (Royal Society 2009, xii, cf. 56, 59)

In regard to CDR approaches, the NRC Committee states: "Public perception of the safety and effectiveness of geological sequestration will likely be a challenge until more projects are underway with an established safety record" (National Research Council 2015a, 67). However, the importance of public perception is even greater in regard to albedo modification.

In relation to both multilateralism and public perception is the idea of international perception of national actions. The NRC report states:

> Moreover, international attitudes towards deployment of albedo modification strategies would have important implications for how any deploying nation or group of people is perceived. Action with even the best intentions can be perceived negatively if those intentions are not clear, and based on demonstrably credible research that supports that such actions would be overwhelmingly positive for humanity. Thus understanding the factors that affect perceptions, and the factors that affect social response to the outcomes of albedo-modification need to be extensively studied in order to strengthen—or at least minimize—the damage to international relationships prior to, during, and post any potential deployment. (National Research Council 2015b, 136)

Perception here is treated as an important factor in communications related to climate engineering actions. Managing perceptions is acknowledged as a legitimate goal in communication as indicated by this explicit and reflective statement. Considerations of public perception management are also explicit elsewhere in the report, for example, when there is discussion of ways in which the scientific community involved in geoengineering might emphasize multiple benefit research to "minimize fears" or transparency and involvement of private contractors (rather than military) to "promote international buy-in and help minimize conspiracy theories" (National Research Council 2015b, 150, 209).

The 2021 NASEM report goes further in explicitly engaging with public perception. In fact, there is an entire section of the report entitled "Public Perception," which includes subsections such as: "Public Acceptance," "Public Trust," and "Framing Considerations" (NASEM 2021, 79–83). Interestingly, the "Framing Considerations" subsection cites many of the same scholars referenced here in making sense of scientific engagement with the public, but rather than analyzing this dynamic from the outside, it is for the purpose of internalizing the findings toward more effective public engagement and public acceptance of a geoengineering research program.

Public perception is indeed a key and recurrent issue throughout the report. As stated by one of the Committee members at the public webinar upon the release of the report: "On the matter of public engagement, the committee is of the view that broad and inclusive participation and engagement are essential if solar geoengineering research is to be viewed as legitimate, useful and deserving of public support" (Al Lin at National Academies Webinar 2021). This idea that transparency and active public engagement are essential elements of legitimacy is recurrent throughout the report. Indeed, the terms "trust and legitimacy" are often used together as a dyad (NASEM 2021, 7, 126, 137, 138, 159). The committee states: "Because SG research and development are controversial and socially consequential, and the technologies themselves could have a range of regional- to global-scale impacts, building trust, legitimacy, accountability, and social responsiveness in both research and research governance are key" (NASEM 2021, 131). (It is worth noting the extent to which the issue of legitimacy is explicitly and directly engaged within this report. Variations of the terms "legitimate" and "legitimizing" are used 46 times in the 2021 NASEM report, up from 5 times in the 2009 Royal Society report, and 4 times in the 2015 NRC report.)

Moreover, the NASEM report deepens and expands the notion of best practices, making issues of justice, diversity, and public engagement central throughout the report. In response to the risk of path dependence, they offer best practices as a line of defense: "Commitments to transparency, justice, and broad engagement in the design and implementation of research will facilitate institutionalization of these values and practices going forward" (NASEM 2021, 6). These issues of ethics and social justice are connected explicitly to the construction of legitimacy: "The integration of research on ethics, justice, and equity into a broader SG research program could enhance both processes and outcomes, strengthening the legitimacy of research and its governance" (NASEM 2021, 242).

Expanding the notion of best practices, this committee deepens the notion of social engagement and appends diversity to the list of best practices, arguing that diversity is important to solar geoengineering research in terms of public engagement including diverse stakeholders and the scientific research itself including diversity of practitioners. In terms of public perception and public engagement, they recognize that: "There is not one monolithic 'public' but rather numerous publics, with different values, worldviews, and perceptions" (NASEM 2021, 79). During the public webinar for the report's release, Committee Chair Chris Field noted:

> A lot of the emphasis on the design of the research program is to make sure that there are pathways for hearing from diverse communities around the world, communities that can express their preferences. [. . .] I don't think there's a way to address the environmental justice issues without listening to vulnerable communities around the world and it will be a critical part of the research moving forward. (National Academies Webinar 2021)

While the NASEM Committee made considerable effort to engage notions of equity, ethics, justice, and meaningful public participation, it should be noted that even so there is a tendency at times to revert to treating favorable public opinion as something to be achieved through strategic efforts. This was noted above in regard to managing public "concerns about a slippery slope toward deployment" (NASEM 2021, 251). This sentiment is especially pronounced in the report when discussing outdoor experimentation. For instance, the report suggests that "moving too quickly and ambitiously toward outdoor experimentation could induce public objections and subsequent delays or restrictions" (NASEM 2021, 251). In this way, the public is treated as a possible challenge to advancing geoengineering research should public opinions and perception not be carefully managed. Similarly, in discussing the development of specific technologies that would be required for deployment, it is stated that there are "two reasons" to wait:

> First, developing detailed designs for deployment now would justifiably raise public concern that the research program was going beyond its stated purpose to solely inform future decisions about deployment. Second, detailed designs are premature, given that many technical requirements will depend on the outcomes of research. (NASEM 2021, 232)

Notably, the first reason stated is due to risk of "public concern" with respect to mission overreach and jumping the gun on deployment readiness as opposed to simply the principled position that premature deployment readiness and technological lock-in should be avoided. These types of references that imply public opinion are something to be carefully cultivated stand antithetical to the explicit ethos presented in the report holding that diverse publics are to be meaningfully engaged and not simply managed in the old "deficit hypothesis" style of public engagement.

In regard to the research community envisioned to participate in geoengineering research going forward, the NASEM report makes the case for diversity on multiple fronts, including geographic representation, interdisciplinary representation, and demographic diversity. They state:

> Diversity is needed in terms of the sites of production of knowledge and the expertise assembled to engage in research. A program will be most effective if it is ambitiously inclusive and systematically incorporates a diversity of stakeholder and disciplinary perspectives, especially those that are typically marginalized. (NASEM 2021, 142)

Arguing for the importance of disciplinary diversity, they state: "There is in particular a need for greater transdisciplinary integration in research, especially linking the physical, social, and ethical dimensions and inclusive of robust public engagement, as well as a need for expanding diversity among the research community itself" (NASEM 2021, 90). With how deeply integrated the issues of justice, transparency, and diversity are in the 2021 NASEM report, this particular report represents a more reflexive engagement with attempting to reconcile social dilemmas and concerns with the simultaneous promotion of geoengineering research.

The Committee's Recommendation 4.1 encapsulates many of the elements of discursive argumentation for the advancement of geoengineering research while holding in mind the multitude of caveats and best practices necessary for avoiding risks of technological lock-in:

> Recommendation 4.1: The United States should implement a robust portfolio of climate mitigation and adaptation. In addition, given the urgency of climate change concerns and the need for a full understanding of possible response options, the U.S. federal government should establish—in coordination with other countries—a transdisciplinary, SG research program. This program should be a minor part of the overall U.S. research program related to responding to climate change. The program should focus on developing policy-relevant knowledge, rather than advancing a path for deployment, and the program should be subject to robust governance. (NASEM 2021, 145)

The various elements within this recommendation include and build upon many of the discursive conventions discussed in this section, including the notion of a portfolio of options (with continued emphasis on mitigation), yet advocating for solar geoengineering research in light of "urgency of climate change" and need to understand all "possible response options" (i.e., Plan B), that the research be multilateral (and transdisciplinary, which is a new element the report added to the discourse), and certainly differentiating research from deployment.

SCIENTIFIC EXPERIMENTS AND PUBLIC COMMUNICATION: THE CASE OF SCOPEX

A demonstrative case study for understanding how the scientific community has evolved in terms of these themes of communication, transparency, and engagement with the public is the SCoPEx research project. An acronym for Stratospheric Controlled Perturbation Experiment, SCoPEx is an open-air experiment being planned by a research team at Harvard University, headed by Frank Keutsch and including other influential figures from the geoengineering community such as David Keith among others. As described by the team, "SCoPEx is a scientific experiment to advance understanding of stratospheric aerosols that could be relevant to solar geoengineering" by doing field testing of what, so far, has only been modeled (Keutsch Group at Harvard 2020). The project website stated in 2020 that a "platform test" is scheduled for June 2021 in Sweden as a step toward an eventual experiment using a high-altitude balloon that will release particles in the stratosphere, allowing for "quantitative measurements of aspects of the aerosol microphysics and atmospheric chemistry" (Keutsch Group at Harvard 2020). However, in March 2021 the project's Advisory Committee recommended postponing "the platform test until a more thorough societal engagement process can be conducted to address issues related to solar geoengineering research in Sweden" (Keutsch Group at Harvard 2021).

The way that this project is presented to the public is particularly notable in terms of the evolution of discursive trends around geoengineering and their direct connection to material developments and activities within the field. The scientists involved in the SCoPEx project have been incredibly cautious, careful, and deliberate in how they communicate to the public. Their careful presentation and explicit transparency seem to reflect and internalize lessons learned in the aftermath of the failure of the SPICE experiment in the UK. As discussed, the SPICE experiment, which would have sprayed water droplets from a balloon distribution platform, was not expected to be controversial and yet significant opposition coalesced around it, ultimately leading to the cancelation of the planned experiment (Cressey 2012; Lukacs 2012; Specter 2012; Owen 2014; NASEM 2021). A significant part of this was the sense that there was a lack of transparency. The organizers of SCoPEx seem to have thoroughly internalized the lessons of SPICE with concerted efforts of transparency, a robust public engagement process, and the formation of an advisory committee for oversight.

Along with other outreach efforts, the research group has maintained a website with a description and presentation of the proposed experiment as well as FAQs for the public. The opening sentence of the webpage describes SCoPEx as "a scientific experiment to advance understanding of stratospheric aerosols that could be relevant to solar geoengineering" (Keutsch Group at Harvard 2020). Notable discursively is that the project is labeled a "scientific experiment" rather than a test. This language keeps much closer to the implication of basic science. Even more noticeable is the hedging, that SCoPEx, a project designed specifically to better understand the effects of solar geoengineering techniques, is described as something "that could be relevant to solar geoengineering" (Keutsch Group at Harvard 2020). These word choices display a proactive discursive strategy to align the SCoPEx project as closely to basic science and multiple benefit research as possible, despite the fact that it is a project with a primary purpose directly related to advancing the understanding of solar geoengineering capabilities. The experiment description goes on to assert: "It is not a test of solar geoengineering per se. Instead, it will observe how particles interact with one another, with the background stratospheric air, and with solar and infrared radiation" (Keutsch Group at Harvard 2020).

The FAQs that follow include a number of opportunities for the team to frame the experiment as a standard research experiment with limited risks in itself. One listed frequently asked question is: "Do other environmental science experiments release materials outdoors?" While it is unclear whether this is a question the public really asks particularly frequently, it does provide the SCoPEx team with an opportunity to reassure the public that there is nothing new in the release of materials outdoors by way of scientific experiment. Similarly, the question is posed: "Is this material dangerous?" providing an opportunity to state: "The test will pose no significant hazard to people or the environment" (Keutsch Group at Harvard 2020).

Another "FAQ" asks: "Will SCoPEx test geoengineering itself?" To this, the public is assured, "This is an experiment not a test." While they state that tests may be important at later stages, they emphasize that this is not the goal of SCoPEx. Rather, it is stated:

This is a science experiment that will (we hope) improve knowledge of some aspects of stratospheric aerosol physics and chemistry relevant to solar geoengineering. This knowledge will improve large-scale models (which are all ultimately dependent on physical observations) that will in turn improve estimates of the overall efficacy and risks of solar geoengineering. (Keutsch Group at Harvard 2020)

In addition to their webpage with an introduction to the experiment and FAQs, SCoPEx has proactively sought public engagement. Anyone interested in the experiment has been invited to sign up to receive email updates on the project. On August 3, 2020, the team published a "Proposed Engagement Process for SCoPEx," and public comment was invited through October 6, 2020. As stated by the principal investigator of SCoPEx, Fred Keutsch, in an email to the interested public:

> We understand that there are complex societal and governance issues surrounding solar geoengineering and SCoPEx. It's why we have taken extraordinary steps to try to perform the experiment in a manner that exemplifies good governance, implementing an unprecedented level of transparency and recruiting a highly qualified, independent, external committee to provide advice that we have and will continue to take very seriously. (Keutsch 2020)

Explaining the public engagement element, he goes on to say:

> We all realize this is a great learning process, so we and the committee are doing everything possible to make sure there are opportunities for people to provide comments and constructive criticism. Such input will help the committee be in the best position possible to weigh whether or not we should proceed with the experiment. We therefore encourage you to provide feedback on their proposed Societal Engagement Process. (Keutsch 2020)

What is fascinating is the extreme or "extraordinary steps" they have been taking in advance of an experiment that objectively does not have significant risks in and of itself. However, they are operating in the post-SPICE environment and taking these extraordinary measures to ensure that the public is on board.

Integrating another lesson learned from the SPICE controversy regarding financial interests and the need for transparency, the project also emailed updates in November 2020 upon the completion of "a review of the project's funding sources to ensure transparency and public disclosure of all funding information" (SCoPEx Advisory Committee 2020a; SCoPEx Advisory Committee 2020b). Again, it is notable that the project team reflexively and explicitly repeats that they are providing "transparency and public disclosure," in effect, proactively staving off any potential accusations to the contrary, like those which have disrupted other projects. The update email directed those interested to a "Financial Review" webpage with four separate PDF files linked that contain both the questions and answers asked by the advisory committee to the SCoPEx Research Team. Despite the assurances of "public disclosure," funding sources are only selectively shared. David Keith writes: "The Statement and Appendix A can be made public at your discretion. Appendix B can be shared privately with the SCoPEx Advisory Committee with the agreement that the information not be released publicly" (Keith 2020). The SCoPEx project has clearly internalized and operationalized best practices for public communication, engagement, and transparency. Even with these proactive and thorough efforts, however, obstacles remain in realizing the goal of entirely transparent public disclosure.

DISCUSSION AND CONCLUSION

Within fields of emerging or contested technologies, discursive representation is paramount to public and political reception. Scientists within such fields, including genetic modification, nanotechnologies and geoengineering, have increasingly recognized the importance of "public consent" (Healey 2014). Discursive strategies can help to present the technology to the public in a more understandable and accessible manner. Moreover, conventionalized discursive strategies can legitimize, rationalize, and bolster public acquiescence and political support of a technological pursuit as was seen in the case of nuclear policy during the Cold War. In like manner, discursive strategies can construct a sense of legitimacy around the pursuit of geoengineering and bolster public acceptance of geoengineering research. This section has identified and examined discursive strategies recurrent in geoengineering science policy reports, which contribute to normalizing the novel, reframing risks, and constructing notions of legitimacy through the privileged "voice of science" (Mukerji 1990), especially a "univocal" voice emanating from a highly respected scientific institution that carefully curates its own projection of legitimacy (Hilgartner 2000). Analysis of the most influential geoengineering policy reports reveals the persistence and, in some instances, deepening of particular discursive strategies over time.

The NRC and NASEM *Reflecting Sunlight* reports follow the precedent of the Royal Society by prefacing each report with the caveat that geoengineering is not a substitute for traditional mitigation efforts and then proceeding with the central message being the need for additional research on geoengineering methods, including those methods of which they caution against deployment for the foreseeable future. As discussed elsewhere, the drive for advancing knowledge is both cultural and material. Within the scientific community, there is a professional culture shaped by the core of the scientific process: inquiry, curiosity, and experimentation. There are also material interests that naturally affect both organizations and individual scientists within the field, even just looking at research scientists in the public and academic sectors (as opposed to industries with vested interests, which will be discussed in the conclusion chapter). Such interests for individuals can be status-based or pecuniary, in the form of grants, research funds, merit-based pay, investment, and future job opportunities. For organizations within related fields, policy in support of geoengineering research can advance new lines of research while also supporting larger organizational missions. Organizations such as the National Aeronautics and Space Administration (NASA), the National Oceanic and Atmospheric Administration (NOAA), as well as the National Laboratories, and major research universities are committed to scientific advancement and are already invested in research directly or tangentially related to geoengineering concepts. Increased support for geoengineering-related research can lead to additional project funding and investment in newer generations of research instruments, which ultimately support scientific advancement both within and beyond the realm of geoengineering research (as indicated in discussion of "multiple benefit" research), hence contributing to the organizations' missions of advancing scientific knowledge.

The intersection of scientific, professional, organizational, nationalistic, and personal dynamics interacting with the face value motivations of geoengineering research translates into a complex and layered form of research advocacy. For example, the goal of advancing albedo modification research is argued in layers: (1) for its own sake (in case we need it), (2) for defensive purposes (in case someone else pursues it), and (3) with the underlying argument that much of the research can be defined as "multiple benefit research" that has other benefits that may come from it even if society is not interested in albedo modification. Combining these different strategies to argue for the pursuit of advancing albedo modification supports a case for advancing research detached from reliance upon any particular strand of argumentation.

Despite acknowledging risks of "moral hazard" and "lock-in," the drive for advancing scientific knowledge and capability is the dominating theme of these reports. The promotion of research is consistent for both CDR and albedo modification, even as the authors display marked reservations regarding deployment of the latter. The overarching promotion of geoengineering research is undergirded by a multi-pronged construction of legitimacy through discursive strategies, including the relative legitimation of actors and approaches, differentiating research from deployment, elevating particular methods through comparative evaluation, and normalizing and naturalizing geoengineering proposals. Together, these discursive trends, recurrent in the most seminal geoengineering policy reports, contribute to constructing a sense of legitimacy for the pursuit of geoengineering.

First, by establishing a hierarchy of legitimacy, especially through constructing the rogue as an unidentified other, notions of risk are reoriented from *what* to *who* and *how*. Within a field defined by risk, this reconceptualization affects perceptions about geoengineering and its trajectory. The 2015 NRC report especially implicates a legitimacy hierarchy of actors and approaches. Emphasizing the potential of unilateral pursuit is used to argue for the necessary legitimate pursuit of albedo modification by the United States in cooperation with other international actors. Like the nuclear state analogy, the convention of establishing a delineation between legitimate and illegitimate actors upholds the internal logic of domestic pursuit of the technology.

Second, furthering the sense of relative legitimacy, certain geoengineering proposals are treated as more credible than others, highlighted through evaluation by comparison (discussed in the previous chapter). The implication that certain geoengineering proposals are superior to others reorients evaluation from considering inherent desirability versus non-action (cf. Bellamy 2013; Bellamy et al. 2012). Moreover, different relative rankings emerge depending on which assessment criteria are privileged, including conceptions of effectiveness, cost, safety, and timeliness (cf. Stilgoe 2015, 117; Bellamy 2013, 2). The benchmark Royal Society and NRC reports warn against and yet utilize evaluation by comparison. While STS analyses examining earlier reports and appraisals have noted and cautioned against the trend of narrow "internal comparisons between geoengineering options" (Bellamy 2013, 927; see also Bellamy et al. 2012; Stilgoe 2015, 117), this trend was continued, and even deepened in the 2015 NRC report (although less apparent in the subsequent 2021 National Academies report). Evaluation by comparison not only serves to elevate particular proposals but also promotes geoengineering more broadly by reorienting questions of risk and feasibility to relative rather than absolute terms, discursively transforming negative language of risks and challenges to positive language of relative advantages and potential.

Third, the differentiation of research and deployment further constructs a paradigm of legitimate pursuit. The NRC report and later NASEM *Reflecting Sunlight* report deepened the discursive distinction between research and deployment, moving from the Royal Society's conception of research as preliminary to possible deployment toward a proactive decoupling of the concepts. The decoupling of research and deployment can be compared to the discursive convention during the Cold War of emphasizing nuclear amassment for the sake of deterrence, not war (Mehan, Nathanson and Skelly 1990). When this discursive convention was "breached" during the Reagan administration, it invoked a public backlash against long-standing nuclear policy (Mehan, Nathanson and Skelly 1990, 134). In the case of geoengineering, since the public is much more open to research than deployment (Corner et al. 2013, 941), discussion of albedo modification deployment rather than research would be expected to increase public opposition to the technology, creating potential obstacles for both research and deployment. Proceeding toward research on albedo modification (but not deployment "at this time") is put forth by the NRC report as a more credible path than that of seeking deployment, even as much of the research advocated is the same that would be necessary for deployment.

Fourth, normalizing and naturalizing climate modification approaches through analogy contributes toward legitimating geoengineering (as discussed in the previous chapter). Science policy reports acknowledge that completely novel planetary conditions would be experienced if various geoengineering schemes were deployed. However, through analogy to natural processes or mundane environmental pollution, these novel schemes are discursively normalized. An earlier study by Corner et al. found that naturalizing geoengineering proposals dispelled concerns among the public and thus cautioned against future geoengineering publications "describing particular geoengineering technologies as 'natural,' or using direct analogies with natural processes" (2011, 26). They argue that "given the importance that participants attributed to the naturalness of the different technologies described, there is a need to ensure that technologies are not associated with the positive notion of 'naturalness' by analogy if, in fact, they are highly artificial" (Corner, Parkhill and Pidgeon 2011, 26). Yet, as demonstrated in this analysis, the 2015 NRC report particularly expanded upon this precedent of normalizing novel geoengineering schemes through recurrent analogy to natural processes and commonplace experiences.

Together, these discursive strategies construct a sense of legitimacy around the pursuit of geoengineering research, especially as encapsulated within prestigious scientific academy reports representing the intersection of science and policy, the line between description and prescription, what "is" known and what "ought" to be done (Stilgoe 2015, 105). They influence the reception of geoengineering concepts among policy makers and the public with material repercussions as future decisions are made on addressing the climate crisis. For instance, the UK government invested in geoengineering research, as promoted by the Royal Society's 2009 report, including grants to technical research and experimentation, such as the SPICE project. However, while the UK-based SPICE open-air experiment was ultimately canceled after public backlash, the Harvard-based SCoPEx project has internalized many of the lessons from SPICE and made major strides in public framing and unprecedented levels of transparency and room for public feedback (although at the time of writing it was still awaiting clearance to proceed pending sufficient public engagement). In terms of government support in the United States, chapter 6 will discuss how by the end of 2017, geoengineering has become politically popular among Congressional members of the House Science, Space, and Technology Committee. Subsequently, NOAA received funding of $4 million allocated from Congress in 2020 to be applied to solar geo-engineering research (Fialka 2020). The Atmospheric Climate Intervention Research Act (H.R. 5519), under consideration by the House Committee on Science, Space, and Technology, would significantly expand support for geoengineering research within NOAA if passed.[3]

Science is not insulated from broader social relations and does not operate autonomously from relations of governance, politics, and public life (e.g., Nowotny, Scott and Gibbons 2001). Beyond acknowledging the importance of public acceptance of technological endeavors, STS scholars have empha-sized the need for meaningful public deliberation and engagement with new technologies that go beyond instrumental dialogue in order to democratize technological decision-making (e.g., Stirling 2014; Stirling 2008; Corner, Parkhill and Pidgeon 2011; Corner et al. 2013; Bellamy et al. 2013; Bellamy 2013; Fischer 2017; Brown and Michael 2002; Nowotny, Scott and Gibbons 2001). Geoengineering represents such a massive intervention in the Earth system that the need for democratic awareness and debate is at least as press-ing as for any other emerging technology, while also particularly problematic (Szerszynski et al. 2013). Rob Bellamy argues that "different instrumental framings can serve to 'close down' on certain geoengineering proposals" and that "Geoengineering assessments should instead seek to 'open up' option and policy choice" (2013, 2; cf. Stirling 2008). While science policy reports themselves acknowledge the need for transparency, deliberation and debate, the most influential geoengineering science policy reports employ discursive strategies that establish prior control over the terms and boundaries of delib-eration. This analysis of geoengineering policy reports especially illustrates how discursive presentation can promote the legitimization of climate engi-neering research.

NOTES

1. This chapter includes material first published as a journal article (Jacobson 2018).

2. Moreover, in the wake of such public opposition to the SPICE experiment, a subsequent open-air experiment (SCoPEx) being planned by a research team at Harvard has implemented lessons learned from the failed SPICE experiment, as will be discussed later in this chapter.

3. The bill, H.R. 5519—Atmospheric Climate Intervention Research Act, which would expand "the directive of the Office of Oceanic and Atmospheric Research within the National Oceanic and Atmospheric Administration to include climate intervention research, including research regarding the effects of proposed interventions in the stratosphere and in cloud aerosol processes," was introduced by Representative Jerry McNerney to the U.S. House of Representatives in 2019 and the last action at the time of writing was in 2020 when it was forwarded from the Subcommittee on Environment to the full House Committee on Science, Space, and Technology for consideration (McNerney 2019).

Section III

JOURNALISM AND PRESENTING GEOENGINEERING TO THE PUBLIC

Chapter 4

Geoengineering Presented to the Public

Narratives and Trends in News Media, 1991–2016

As of yet, geoengineering exists in concept, rather than in practice. The conceptions of geoengineering are shaped and reshaped within the realm of "socio-technical imaginaries," a term coined by STS scholar Sheila Jasanoff, which she defines as the "collectively held and performed visions of desirable futures [. . .] animated by shared understandings of forms of social life and social order attainable through, and supportive of, advances in science and technology" (Jasanoff 2015, 19; cf. Baskin 2019; Bellamy et al. 2012; Stilgoe 2015, 7–10; Stilgoe 2016; Healey 2014; Markusson 2013, 4; Corner et al. 2013). Geoengineering as a concept (or as a collection of "socio-technical imaginaries") requires translation from the models and theories of scientists to the understanding of policy makers or the general public. While science policy reports (as discussed in the previous section) represent the translation of geoengineering imaginaries into language accessible (and *assessable*) to political decision-makers and relevant stakeholders, news articles by science reporters, journalists, and editorialists bring the imaginaries and possibilities of geoengineering to a broad public audience.

It is primarily through news media and scientific journalism that the concepts of geoengineering are presented to the public. The 2021 NASEM *Reflecting Sunlight* report references the role of media, stating: "The discourse around SG is also shaped by (and in turn shapes) coverage in the media" (NASEM 2021, 83). Even among the subset of the public most interested in the topic, few will ever read the geoengineering reports by the Royal Society or the National Academy of Sciences, but they are likely to read the *Guardian* or the *New York Times* science editors' articles discussing these reports and quoting the expert informants. As such, news media serve as a communication bridge between experts and the public.

This section contributes toward a better understanding of the public discourse on geoengineering and its trajectory through a longitudinal analysis of articles from mainstream English-language news sources from the United States and United Kingdom. The research was structured to include a broader range of news sources than previous studies (e.g., Luokkanen, Huttunen and Hildén 2014, 967), with inclusion of both print and online journalistic reporting and analytical editorials. It also opens up the scope of analysis to be inclusive of both substantive content and sources as well as rhetorical features and discursive framing. A corpus of 94 articles substantially focused on geoengineering was analyzed through close reading and qualitative coding for relevant themes, subjects, actors, and discursive-rhetorical practices.[1] A complementary quantitative analysis confirmed that the trends seen within the qualitative study were consistent with the broader universe of English-language news articles that discuss geoengineering and also provide insight on trends and relative change over time.[2]

News media is a communication genre that not only introduces novel concepts to the public but also provides a proxy measure of public attention to such issues. This chapter will examine the broad trends seen in a quarter century of news coverage of geoengineering. It will present quantitative trends in geoengineering reporting over time, highlight central narratives in the discourse, and examine the prominence of certain voices in how geoengineering is covered in news publications before looping back to the theme of science policy reporting from the previous section, considering how media present such reports into the public sphere. The following chapter (chapter 5) then presents a detailed analysis of specific discursive conventions that guide the conceptualization and assessment of geoengineering within the journalistic genre.

SHIFTING SENTIMENT AND MOVE TO MAINSTREAM

Geoengineering emerged as a theme in popular media in the mid-2000s, with 2009 marking a significant shift in its increased attention. Using the *New York Times* as an example, this influential newspaper's first article on geoengineering was published in 2006, followed by three articles in 2007 (in addition to at least three op-eds or letters to the editor), nine articles in 2008, 20 in 2009, then 10 or less in subsequent years through 2016 (see figure 4.1).[3] For the *New York Times*, geoengineering articles clearly spiked in 2009. Similarly, *Newsweek* magazine published a single article on geoengineering in 1991 and then turned to the subject again starting in 2007 with two articles, followed by one in 2008, and with a spike of eight in 2009 before returning to four or less for subsequent years through 2016. Looking at a non-specialist audience

New York Times ▬▬Scientific American ▬▬Newsweek

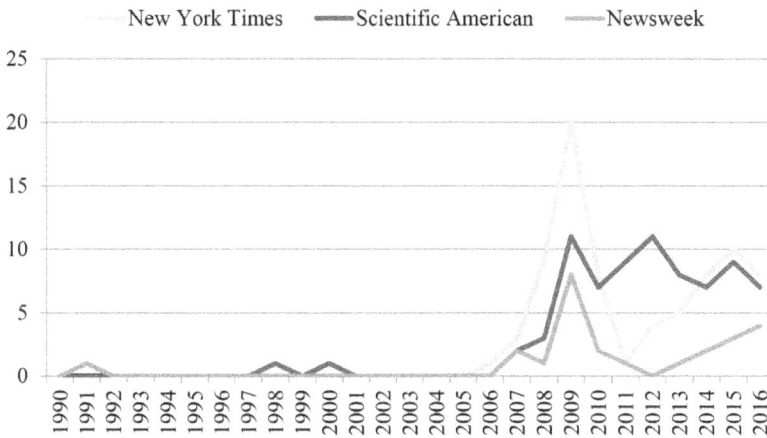

Figure 4.1   News Articles Substantially Discussing Geoengineering for Three Periodicals by Year. *Source*: Created by author.

publication focused on science, the *Scientific American* discussed proposals that fit in the category of CDR in individual articles in 1998 and 2000, then in 2007 two *Scientific American* articles discussed geoengineering, three articles focused on geoengineering in 2008, eleven in 2009, and remaining fairly steady in the subsequent years with at least seven articles annually through 2016. For the *Scientific American*, 2009 also represented the first dramatic rise in articles devoted to geoengineering, with attention in this publication remaining fairly sustained in subsequent years. (See figure 4.1 for a visual representation of these data.)

A similar pattern can be seen in international aggregate news sources. To capture this, figure 4.2 shows by year the total English-language newspaper articles that reference geoengineering from all news sources within the LexisNexis database.[4] Again, the year 2009 shows a dramatic peak with 345 newspaper articles referencing geoengineering. There is a secondary peak in 2015 with 272 relevant articles. These two years with clear spikes in media attention to geoengineering correspond with the publication of important documents within the field, respectively the 2009 UK Royal Society report on geoengineering and the 2015 reports by the U.S. National Research Council of the National Academy of Sciences.

Journalistic coverage of geoengineering is obviously tightly coupled to events and publications within the scientific field. Within the geoengineering field itself, there are certain core publications that have influenced, legitimated, and illuminated the evolving discourse of geoengineering. The three most pivotal are a 2006 essay by Nobel laureate Paul Crutzen encouraging consideration of albedo modification, the 2009 Royal Society Report, and the

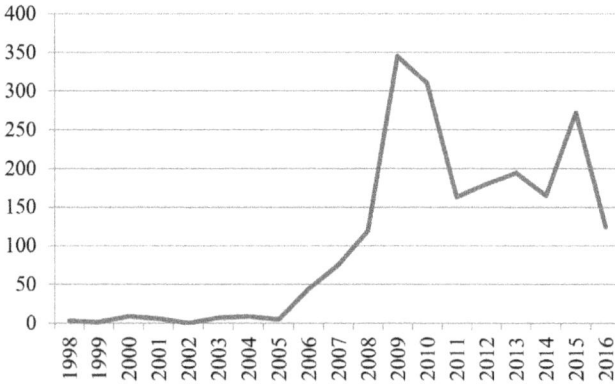

Figure 4.2 Combined English-Language Newspaper Articles That Discuss Geoengineering by Year Listed in the LexisNexis Database. *Source*: Created by author.

2015 NRC report. These publications have influenced the public discourse on geoengineering in both content and quantity, with the number of news articles by year reflecting the release of these seminal documents. Attention to geoengineering within news media began to increase in 2006, coinciding with the publication of Crutzen's seminal article, the significance of which will be discussed in detail below. News coverage of geoengineering peaks to its highest rate in 2009, which is also the publication year for the Royal Society report, and experiences another spike in 2015, the year of the NRC report (see figure 4.2). As discussed in the previous section, these two reports have been the preeminent benchmark science policy reports related to geoengineering, and their treatment of the subject has contributed to increasing a sense of legitimacy and mainstreaming of geoengineering, which is in turn reflected in news media discourse.

In terms of journalistic coverage of geoengineering, 2009 was a pivotal year both quantitatively and qualitatively. First, 2009 represented a major spike in the attention given to geoengineering, as indicated by the peak number of geoengineering articles (see figures 4.1 and 4.2). Second, 2009 also marked the transition point of perceiving geoengineering's move from the fringe toward the mainstream. Looking at use of the word itself, "fringe" appears explicitly within the corpus of articles in the years 2009 and 2010. In 2009 "fringe" is still used in the present tense, and in 2010 usage is split between present and past tense. For example, David Victor et al. write in 2009 that "geoengineering continues to be considered a fringe topic" (Victor et al. 2009, 73). Relatedly a *New York Times* article attributes to David Victor the perspective that geoengineering "needs to be brought in from the mad-scientist fringe" (Lohr 2009). In March 2010, the *Economist* writes that "modern climate scientists [. . .] usually see geoengineering research as niche, if not fringe, stuff" (2010b). However, by December 2010, "fringe" becomes a past tense descriptor as stated by an Associated Press journalist reporting from the Cancun climate talks: "Just a few years ago, geoengineering was regarded as a fringe idea, a science-fiction playground for imaginative scientists and engineers" (Hanley 2010).

In the years after 2010, the term "fringe" largely disappears from the discourse. It is entirely absent from the corpus of news articles reviewed here from 2011 forward. From this point forward, a common theme in popular media literature is the shifting sentiment surrounding geoengineering, particularly its move toward mainstream consideration. This discursive trend includes three elements which may be integrated or stand-alone: (1) emphasizing geoengineering's move from the fringes of science toward mainstream consideration, (2) pointing to certain scientific publications by particularly esteemed individuals or scientific societies as indication of the move toward mainstream, and (3) connecting the increasingly mainstream consideration of geoengineering to the increasingly notable effects of climate change and the problematic state of mitigation efforts.

General audience articles on geoengineering often include these multiple elements of discussing the shift toward mainstream consideration. An exemplary 2014 article by Henry Fountain, climate journalist for the *New York Times*, neatly encapsulates all three of these themes:

> Once considered the stuff of wild-eyed fantasies, such ideas for countering climate change—known as geoengineering solutions, because they intentionally manipulate nature—are now being discussed seriously by scientists. The National Academy of Sciences is expected to issue a report on geoengineering later this year.
>
> That does not mean that such measures, which are considered controversial across the political spectrum, are likely to be adopted anytime soon. But the effects of climate change may become so severe that geoengineering solutions could attract even more serious consideration. (Fountain 2014)

The first sentence of this quotation emphasizes the perception of the fringe origins of geoengineering with the evocative phrase "Once considered the stuff of wild-eyed fantasies" and how such markedly offbeat origins have shifted now to "being discussed seriously by scientists" (Fountain 2014). The unmodified term "scientists" as used here carries with it the legitimacy and respectability of the profession, especially when placed in rhetorical opposition to "wild-eyed fantasies." The legitimacy of scientific consideration is further emphasized by reference to the report underway at the time by the respected and undeniably scientific mainstream National Academy of Sciences. Finally, the author states that, despite the controversy over such methods, "the effects of climate change may become so severe that geoengineering solutions could attract even more serious consideration" implicitly recognizing the insufficiency of mitigation efforts to date as setting the premise for mainstream consideration of geoengineering (Fountain 2014). These three themes encapsulate the sense of shifting sentiment on geoengineering as portrayed in general audience publications and will be considered individually below.

Articulating Fringe Origins and the Move toward Mainstream

Consistently from 2010 through 2016, popular media articles on geoengineering emphasize its move from the fringes of science toward the mainstream. According to a 2010 article addressing the emergence of geoengineering as a potential consideration at the Cancun international climate talks:

> "The taboo is broken," Paul Crutzen, a Nobel Prize-winning atmospheric scientist, told The Associated Press. Whatever the doubts, "we are amazingly farther up the road on geoengineering," Crutzen, who wrote a 2006 scientific article that sparked interest in geoengineering, said. (Hanley 2010)

The notion that there once was a "taboo" regarding the open discussion of geoengineering, which has subsequently been broken, is a recurring theme in popular discourse on geoengineering, reflecting a perspective articulated among geoengineering advocates. As will be discussed in greater detail, the essay referenced in the quotation above, written by Nobel laureate chemist Paul Crutzen and published in the journal *Climatic Change*, is consistently identified as a critical factor in breaking the "taboo" on open discussion of geoengineering. As such, it became a core element in the dominant narrative around the progression of geoengineering as conveyed and reinforced within news media.

Similar to the idea of a taboo that has been removed is the theme of fringe science versus mainstream science and the significant shift of geoengineering from the former to the latter category in both practice and perception. For instance, the exemplary *New York Times* article by Henry Fountain, mentioned earlier, juxtaposes the reaction over time to geochemist Olaf Schuiling's advocacy for using olivine (a mineral that absorbs carbon dioxide) as a form of CDR:

> "When I started, I was a nutty professor," Dr. Schuiling said. But when he gives a talk nowadays, "the first question after I finish is, 'Why don't we do it?'" (Fountain 2014)

Here the shift in perception among the audience is made abundantly clear by highlighting how the same scientist with the same message went from being dismissed to being taken seriously.

The *shift to mainstream* narrative both reflects and is reinforced by the viewpoints of individuals interviewed and quoted by journalists. During 2009, the key turning point year in transitioning geoengineering discourse, one reporter cites David Victor as saying: "Most analysts who examined the options closely had concluded that it would be reckless to mess with the planet. [. . .] That is changing" (cited in Fischer 2009). The taboo narrative as well as the related, although somewhat contradictory, fringe-to-mainstream narrative have become entrenched in geoengineering discourse, with geoengineering advocates variously promoting both narratives.

While the narrative of fringe origins has been prevalent within geoengineering discourse, its premise is dependent upon where the starting point in the timeline of geoengineering's intellectual history is set. As discussed in the introduction chapter, the origins of the concept of geoengineering in response to climate change go back to the 1965 Report of the Environmental Pollution Panel of the President's Science Advisory Committee. Again, the concept surfaces in the 1992 NRC report (as will be discussed later in this chapter). This presumably leaves the period of time from the 1992 NRC report through 2006 when the "taboo" was "broken" by Crutzen's article as the dark ages for geoengineering. The fringe origins narrative is seemingly dependent, then, on setting the origins of consideration in this period. Furthermore, although they serve similar purposes in emphasizing the move to mainstream, there is some contradiction between the fringe narrative and the taboo narrative. The taboo narrative implies that mainstream scientists were silenced in regard to geoengineering, while the fringe narrative obfuscates the relationship of mainstream scientists with geoengineering until the point at which it is considered to have moved mainstream.

In any case, as the "taboo" has lifted since 2006, experts and proponents have emerged willing to discuss geoengineering. Frequent interviewees include core members of the geoengineering community. Science journalist Eli Kintisch in his book *Hack the Planet* (2010, 8) coined the term "Geoclique," which subsequently Clive Hamilton expanded upon in his book *Earthmasters* to refer to the "constituency for geoengineering [. . .] developing around a network of individuals with personal, institutional and financial links" (Hamilton 2013a). Hamilton specifies: "At the centre of the network is a pair of North American scientists actively engaged in geoengineering research—David Keith and Ken Caldeira" who "have been dominant voices in virtually every inquiry into or report on geoengineering" (Hamilton 2013a). The present research reaffirms that Keith and Caldeira continue to be "dominant voices" and not only internally within the geoengineering community but as the voices communicating it to the public through citation in news reports and editorials.[5]

David Keith in particular is disproportionately cited compared to other experts within the field. For at least 10 years, Keith was consistently the most cited scientist, by far, within public discourse in regard to geoengineering. Within the sample of articles studied here that are substantially focused on geoengineering, 28% cite to David Keith (see table 4.1). Looking at the broader universe of articles with any reference to geoengineering, over 5.4% cite to David Keith in the years 2007–2016 (see table 4.2). The next most cited individual, Ken Caldeira, is cited in 16% of the sample and 4% of the broader 10-year period sample from LexisNexis. For comparison, the chair of the National Academy of Sciences/National Research Council Committee on Geoengineering, Marcia McNutt, garnered reference in 5.4% of the sampled articles and 0.8% of articles referencing geoengineering within the 10-year period in the LexisNexis database.[6]

Among the frequently cited individuals in the broader search, Bjørn Lomborg stands out in the fourth position behind David Keith, Ken Caldeira, and Paul Crutzen in table 4.2, showing percentage of articles citing to these individuals among those discussing geoengineering in the LexisNexis database. Lomborg proclaimed himself "The Skeptical Environmentalist" with his 2001 book of this title. He has been a controversial figure in relation to

Table 4.1 Most Frequently Cited or Referenced Individuals in Geoengineering-focused Articles

Person	# Articles Referencing	% Articles Referencing
David Keith	26	27.7%
Ken Caldeira	15	16.0%
Alan Robock	11	11.7%
Paul Crutzen	11	11.7%
Oliver Morton	8	8.5%
Raymond Pierrehumbert	8	8.5%
Clive Hamilton	7	7.4%
John Latham	7	7.4%
David Victor	6	6.4%
Jane Long	5	5.3%
Marcia McNutt	5	5.3%
Edward Teller	5	5.3%
Hugh Hunt	4	4.3%
Eli Kintisch	4	4.3%
Stephen Salter	4	4.3%
Victor Smetacek	3	3.2%

Note: Based on sample of articles focused specifically on geoengineering. $N = 94$

climate change as he has gone against the grain of overwhelming consensus on the need to pursue mitigation to address anthropogenic climate change. He is known as a global warming skeptic (Weisenthal 2009), acknowledging existence of anthropogenic climate change but downplaying its risks and urgency to act. For example, he openly campaigned against the Kyoto Protocol (Dasgupta 2007). In terms of geoengineering, Lomborg is emblematic of climate skeptic turned geoengineering advocate. This category also includes organizations such as American right-wing think tanks like the American Enterprise Institute, the Climate Response Fund, and the Climate Institute (Sikka 2012, 163–164). Lomborg also stands out as the most cited non-scientist on the list, which is particularly striking given what a controversial figure he is. (See table 4.2 for demarcation by category of which of the frequently cited individuals are scientists, engineers, or social scientists.) As will be discussed further, controversy is a recurrent theme emphasized by news media, and inclusion of controversial figures contributes to this.

Table 4.2 includes three columns of values, two five-year periods and the aggregate 10-year period of 2007–2016. This differentiation gives a sense of which individuals remain consistently influential in news media over the 10-year period and those whose voices in public discourse increased or

Table 4.2   Most Frequently Cited or Referenced Individuals in Articles Discussing Geoengineering, 2007–2016 and 2012–2016 Compared, and Total 2007–2016

	2007–2011	*2012–2016*	*2007–2016*
David Keith*	5.2%	5.7%	5.4%
Ken Caldeira*	4.7%	3.2%	4.0%
Paul Crutzen*	4.9%	1.9%	3.5%
Bjørn Lomborg^	4.4%	2.2%	3.4%
Alan Robock*	1.8%	2.5%	2.1%
Clive Hamilton^	0.5%	3.0%	1.7%
Stephen Salter**	2.8%	0.4%	1.6%
John Latham*	1.9%	0.6%	1.3%
Marcia McNutt*	0.0%	1.7%	0.8%
Raymond Pierrehumbert*	0.3%	1.3%	0.8%
Hugh Hunt**	1.2%	0.1%	0.7%
Eli Kintisch^	0.7%	0.5%	0.6%
Victor Smetacek*	0.7%	0.5%	0.6%
David Victor^	0.8%	0.2%	0.5%
Jane Long**	0.3%	0.5%	0.4%

Note: Based on sample of articles referencing geoengineering from LexisNexis database (controlled for high similarity), 2007–2011 (*N* = 1,012), 2012–2016 (*N* = 935), and 2007–2016 (*N* = 1,947)
* = Scientist (e.g., climatologist, physicist, atmospheric scientist)
** = Engineer
^ = Social scientist (e.g., political scientist, ethicist, journalist)

decreased during this time. For instance, David Keith is consistently prominent. He is the most cited individual in connection to geoengineering within public discourse in the last 10 years by any measure. In contrast, Marcia McNutt had no public exposure as measured by newspaper citations from 2007 through 2011 but later (in 2015) came to be cited relatively frequently due to her role as chair of the NRC Committee responsible for the pivotal 2015 geoengineering report. Other individuals such as Bjørn Lomborg, Stephen Salter, John Latham, and Hugh Hunt received fewer citations in 2012–2016 relative to their representation in public discourse the previous five years.

Emphasizing Failure of Mitigation to Explain Geoengineering's Move toward Mainstream

A prevalent framing in news media articles about geoengineering, especially within editorials, is emphasizing the poor state of mitigation efforts to explain geoengineering's move toward mainstream. An op-ed in the *New York Times* by Joe Nocera (2015) is a clear example of this framing, which involves a flow of three steps: (1) emissions controls are identified as the best answer to confronting climate change, (2) their failure to date is acknowledged, and so (3) it is argued that geoengineering must be considered. In Nocera's editorial it went like this:

> What's the best way to reduce the chances of climate change wreaking havoc on Earth? [. . .] The most obvious answer—one we've known for years now—is to reduce the amount of carbon dioxide we're pumping into the atmosphere. [. . .] Despite this knowledge, however, few policies have been put in place to spur any of that. [. . .] So maybe we need to start thinking about coming at the climate-change problem from a different direction. Instead of hoping that humans will start reducing their carbon use, maybe it's time to at least consider using technology to keep climate change at bay. (Nocera 2015)

The imperative of geoengineering is framed as a direct result of failed or insufficient mitigation efforts.

The framing is sometimes more concise. Brad Plumer writes for the *Washington Post*:

> Many of the world's nations show few signs of cutting their greenhouse gas emissions anytime soon. That's why, in recent years, more and more climate scientists have been pondering the concept of "geoengineering" as a way to slow the pace of global warming. (Plumer 2014)

Here, the problem is identified as emissions, and the failure of cutting emissions is used to explain the shift toward mainstream scientific consideration of geoengineering.

A 2009 article published in *Scientific American* opens with the premise that insufficient mitigation efforts have opened the door to geoengineering proposals, stating: "Failure to make difficult choices to cut greenhouse gas emissions exposes humanity to an increasingly dire set of climate scenarios. But there is a way to buy time: Geoengineering" (Fischer 2009). The author then discusses the controversial nature of geoengineering and the range of geoengineering options, all of which "have major drawbacks" (Fischer 2009). The article concludes nonetheless that, despite the controversy and drawbacks, "the concept is gaining more traction as politicians, confronted with the ugly reality of trying to wean economies off fossil fuels, cast about for a strategy that will work if climate changes quickly or in nasty ways" (Fischer 2009). Fischer quotes David Victor in saying that "most analysts who examined the options closely had concluded that it would be reckless to mess with the planet. [. . .] That is changing." This changing sentiment on considering geoengineering is explained as having transitioned from being characterized as recklessness due to risks of moral hazard threatening abatement efforts to the point of mainstream consideration largely due to the emerging consequences of climate change in the face of insufficient mitigation efforts. As Fischer asserts:

> It's changing, in large part, because the chances of any sort of international agreement on radical emissions cuts are plummeting even as scientists find evidence that these emissions have the potential to destabilize the Earth's climate to a degree unforeseen in human history.
>
> If those predictions come true, scientists fear any hand-wringing over the consequences of planet-wide mitigation will pale in comparison to the inconsolable pleas of populations facing rising seas, searing dust storms and savage famines, scientists warn. (Fischer 2009)

The implication is that, despite the taboo on geoengineering, insufficient mitigation efforts have paved the way for geoengineering proposals to garner more serious attention.

Drawing upon influential voices in geoengineering discourse, Michael Specter demonstrates the logic used by research advocates in propelling geoengineering progress. He quotes David Keith as saying: "There will be no easy victories, but at some point we are going to have to take the facts seriously" (Keith quoted in Specter 2012). Citing Crutzen, Specter goes on to explain:

Although the I.P.C.C., along with scores of other scientific bodies, has declared that the warming of the earth is unequivocal, few countries have demonstrated the political will required to act. [. . .] With each passing year, goals become exponentially harder to reach, and global reductions along the lines suggested by the I.P.C.C. seem more like a "pious wish," to use the words of the Dutch chemist Paul Crutzen. (Specter 2012)

To clinch the argument, he then turns to Lord Rees:

"Most nations now recognize the need to shift to a low-carbon economy, and nothing should divert us from the main priority of reducing global greenhouse gas emissions," Lord Rees of Ludlow wrote in his 2009 forward to a highly influential report on geoengineering released by the Royal Society, Britain's national academy of sciences. "But if such reductions achieve too little, too late, there will surely be pressure to consider a 'plan B'—to seek ways to counteract climatic effects of green-house gas emissions." (Specter 2012)

Again, the implication of this course of argument is that geoengineering must be taken seriously as an option to address climate change as a direct result of mitigation not being taken seriously by policy makers.

Discussion of geoengineering at the 2010 Cancun climate talks marked an important moment in the mainstreaming of geoengineering discourse. Reporting from the talks, one journalist states:

Like the warming atmosphere above, a once-taboo idea hangs over the slow, frustrating U.N. talks to curb climate change: the idea to tinker with the atmosphere or the planet itself, pollute the skies to ward off the sun, fill the oceans with gas-eating plankton, do whatever it takes. (Hanley 2010)

This observation couples the increasingly notable effects of climate change with the newfound openness to broach geoengineering as a serious option rather than focusing climate negotiations solely on abatement.

The taboo narrative is premised on two concerns regarding consideration of geoengineering: the moral hazard concerns that pursuit of geoengineering may reduce GHG abatement efforts, as discussed in the introduction, and the direct risks and potential for incalculable ecological harm that could result from geoengineering deployment. Sharon Begley, who has written a number of articles on geoengineering for *Newsweek*, showcases the point that geoengineering is increasingly considered *despite its serious risks*. In one article, she draws upon a publication entitled "20 Reasons Why Geoengineering May Be a Bad Idea" published in the *Bulletin of the Atomic Scientists* by Alan

Robock, a wary geoengineering researcher often quoted for expert commentary (see tables 4.1 and 4.2):

> After decades spent studying volcanoes, Alan Robock can list 20 reasons why humans should not try to play God with the world's climate by [deploying albedo modification with stratospheric sulfuric aerosols, which] might counter the global warming caused by carbon dioxide and other greenhouse gases. But that's not all sulfates do [. . .]
>
> The particles also deplete the planet's ozone layer, which is just starting to repair itself now that ozone-shredding chemicals are banned. They cause acid rain, too. And by cooling large land masses like Asia and Africa, the heat-reflecting particles reduce the temperature difference between them and the already-cooler oceans, which could stifle the monsoons that millions of people depend on for agriculture. Because the particles block direct sunlight more than diffuse rays, they also alter the balance of radiation reaching Earth's surface, with unknown consequences for plants that can be kind of finicky about the kind of sunlight they need.
>
> And yet . . . In a sign of how dangerous global warming is starting to look and of how pitiful the world's efforts to control greenhouse gases are, even Robock—list and all—hedges his bets. Geo-engineering, allows the Rutgers University meteorologist, "might be held in reserve for an emergency." (Begley 2007)

This lengthy quote paints a stark contrast between the serious risks associated with SAAM and the fact that even some of the scientists most aware of these risks, and deeply concerned about them, still see Plan B pursuit as necessary. The Plan B in-case-of-climatic-emergency framing of geoengineering is a central strand of narrative throughout geoengineering discourse, including in scientific publications, science-policy reports, and statements, as well as popular media (cf. Bellamy 2013, 1; Bellamy et al. 2012, 605, 609; Nerlich and Jaspal 2012, 142; Corner, Parkhill and Pidgeon 2011, 13; Corner et al. 2013, 945). The narrative is premised on the idea that there is some critical (although currently undefined) tipping point at which the dangers of climate change may be determined to outweigh the dangers of geoengineering. Begley's juxtaposition of the dangers of geoengineering enumerated by Robock paired with his reluctant position that it may prove necessary anyway highlights the seriousness of unmitigated climate change by showcasing the extreme risks of geoengineering solutions being considered to address it.

After the Paris international climate summit in 2015, which is considered to be the most successful summit to date, geoengineering again emerged within popular discourse as a potential outcome of that summit, which officially only focused on mitigation as have all other UN climate talks to date. An editorial by a legal scholar asserts that the Paris Accord

> establishes an aspiration goal of holding climate change to 1.5°C, with a firmer goal of holding the global temperature decrease "well below" 2°C. As a practical matter, the 1.5°C goal almost certainly would require geoengineering, such as injecting aerosols into the stratosphere or solar mirrors. (Farber 2015)

Similarly, an article in *Slate* points to pursuit of geoengineering as an underlying risk to post-Paris climate dynamics, despite the lofty emission goals set and the absence of geoengineering from the official discourse of the summit: "The historic agreement forged in Paris among 195 countries in December holds the promise of triggering a global shift to combat climate change—and harbors a hidden warning" in regard to the possibility of geoengineering (Venkataraman 2016).

The author asserts that while "the Paris accord is a triumph of diplomacy," its "success in heading off the worst climate disruptions hinges on whether countries fulfill the pledges each made leading up to the Paris talks and make bolder ones this decade" (Venkataraman 2016). However:

> The United States faces strong internal pressure to keep burning fossil fuels, reflected in our divisive politics; other nations—especially island nations like Tuvalu and Kiribati—face strong pressure to keep the planet cooler at any cost. The seas are already rising. (Venkataraman 2016)

From the contradiction of these antithetical national interests, the author pivots to the prospect of geoengineering:

> The mood is ripe for private-sector companies or individual nations to seek drastic ways to change the climate, either to avoid the cuts agreed to in Paris or to hedge their bets in case of political failure. Yet absent from the Paris agreement and absent from U.S. political discourse is any robust discussion of what could be a growing threat, especially after the November presidential election: that countries, people, or businesses will take it upon themselves to directly cool the planet. (Venkataraman 2016)

This line of speculation culminates with the simple factual statement: "Experiments in geoengineering have already been tried" (Venkataraman 2016). Given the material manifestations of climate change and differing national interests, the indication is that unilateral geoengineering is an underlying risk should the spirit of Paris fall short of realizing bold abatement efforts. This editorial expressing concern regarding the possibility of geoengineering adopts a form of the climate emergency narrative, premising the risk of geoengineering, especially done unilaterally (another prevalent theme in geoengineering discourse as discussed in the previous chapter), on the failure of the global community to adopt sufficient carbon abatement measures leading individual entities to subjectively determine that the emergency or catastrophe threshold has been reached such as to justify geoengineering deployment.

Even with the relative success of the Paris Summit, its contextual position as one in a long line of historically failed agreement efforts provides an opening argument to advance geoengineering due to failure of achieving meaningful mitigation policy.

> As negotiators at the climate talks underway here spar over what to do about adding more CO_2 to the air, geoengineering becomes more and more attractive to those with this tinkerer's bent [. . .]
>
> The incredibly slow progress in combating climate change worldwide—the Paris talks are the 21st attempt to reach international agreements in the past 25 years—raises the appeal of the seemingly quick fix of seeding the sky. I remember attending a panel on geoengineering with Morton back in the heady days before the 2009 United Nations Climate Change Conference negotiations in Copenhagen. As Morton and his fellow panelists pointed out, with little hope to cut pollution, artificial volcanoes or a fleet of aircraft spewing out sulfur might prove not just enticing but necessary. (Biello 2015)

This selection is illustrative of a theme within public discourse on geoengineering that indicates a sense of inevitability arising from the failure of mitigation negotiations. The public or political acceptance of inevitability, of course, can have material effects in the trajectory of the field. An entrenched sense of inevitability is akin to the social psychology concept of a "self-fulfilling prophecy" that "itself produces the requisite conditions for the occurrence of the expected event" (Watzlawick 2011 [1984], 393).

Using Scientific Publications as Indication of Move to Mainstream

A chapter on geoengineering was included as an area of possible consideration in the NRC Committee on Science, Engineering, and Public Policy 1992 publication "Policy Implications of Greenhouse Warming: Mitigation, Adaptation, and the Science Base." This inclusion did not garner much attention compared to the subsequent publications focused on geoengineering. However, at the time, science journalist Sharon Begley homed in on the significance of the NRC's treatment of geoengineering, foreshadowing many themes that would be seen more than two decades later when the NRC reexamined geoengineering in its extensive 2015 report on the subject. After emphasizing the "zaniness that has kept [geoengineering proposals] out of the scientific mainstream," she points to the inclusion of geoengineering in what was the forthcoming 1992 NRC report as indication of its move toward the scientific center:

> But now these schemes may be ready for their day in the sun. In a soon-to-be-released report, the National Research Council (NRC)—the operating arm of the prestigious National Academy of Sciences—endorses further study of geoengineering, granting the field a legitimacy it has so far lacked. Although the panel does not support even pilot programs, it calls geoengineering "technically feasible in terms of cooling effects and costs" and says it has "the potential to affect greenhouse warming on a substantial scale." (Begley 1991)

As it turned out, the schemes were not quite yet "ready for their day in the sun" and after the 1992 publication, there was a significant gap during which geoengineering did not receive mainstream scientific attention. As mentioned, it is often said that geoengineering became a taboo subject during this time period as not to interfere with the scientific community's emphasis on emissions abatement in response to climate change. This changed in the early 2000s. Publications from esteemed scientists, such as Paul Crutzen's seminal 2006 article breaking the "taboo" on advocating pursuit of albedo modification, and reports from esteemed scientific organizations, notably the Royal Society's 2009 report and the NRC's 2015 report, have served as indication of geoengineering's move to the mainstream as interpreted by popular media. These publications have significantly influenced the trajectory and discourse of geoengineering both within the relevant scientific communities and also externally, bridging the technical fields and public discourse around them.

In the *Scientific American*, Douglas Fischer explains the significance of Crutzen's article breaking the taboo on openly discussing to possibility of albedo modification:

> For years [. . .] it was taboo [to discuss geoengineering] on the fear that, if climate control was seen as a viable option, pressure on world leaders to reduce emissions might ease. [. . .] That changed in 2006 with the publication of a seminal essay in the journal *Climatic Change* by Nobel laureate Paul Crutzen, emeritus professor at the Institute for Marine and Atmospheric Systems at Utrecht University in the Netherlands. (Fischer 2009)

Crutzen's article was groundbreaking as an open endorsement of geoengineering pursuit by a prominent scientist, notably a Nobel laureate with all the legitimacy that title entails, in a mainstream scientific journal that had previously, like others of its ilk, steered clear of geoengineering. As reported in a contemporaneous *New York Times* article, the publication of Crutzen's piece was not taken lightly but rather imbued with controversy, characterized as a "bitter dispute" (Broad 2006). After significant negotiation, Crutzen's paper was ultimately published as part of a "compromise" in which a number of commentaries on the topic of geoengineering, from multiple perspectives, were included in the same issue presumably to provide a balance of perspectives and offset the impacts of Crutzen's argument (Broad 2006).

Despite this compromise, however, it is clearly Crutzen's piece that had the most lasting impact in subsequent years, greatly influencing geoengineering discourse within scientific communities and among the public. It is frequently referenced as a critical point in breaking the "taboo" on geoengineering. For example, a book review in the *Economist* of contemporary publications on geoengineering credits Crutzen for the turning point in geoengineering consideration:

> In 2006, depressed by the lack of progress on emissions, Paul Crutzen, an atmospheric researcher, broke a long-standing taboo among climate scientists by publicly pointing out that if humans have the power to heat the planet, then they also have the power to cool it down again. (*The Economist* staff 2013)

A *Slate* article on the who's who of geoengineering describes Paul Crutzen as follows: "Crutzen, a Nobel Prize-winning atmospheric chemist, helped legitimize scientific conversations about geoengineering with his 2006 paper about seeding the atmosphere with sulfur to reflect sunlight back into space" (Brogan 2016). These examples, as well as the Fischer (2009) citation above, credit the Crutzen article with breaking the "taboo" on mainstream geoengineering discussion and notably point to its role in legitimizing the field.

More than simply breaking the taboo, Crutzen's article marks an important transition point in the legitimacy of the field. It paved the way for other mainstream publications on geoengineering and influenced the subsequent publication of scientific reports that have added a new layer of perceived legitimacy to the field of geoengineering. In the introduction chapter to the NRC report, the most extensive scientific report on geoengineering to date, the subsection entitled "Motivation for researching albedo modification" explicitly cites to Crutzen's article in its explanation, stating: "Crutzen (2006) raised the question of whether humanity might want to develop the capability [of] intentionally modifying Earth's albedo to a greater degree and offset a larger amount of forcing" (National Research Council 2015b, 31).

Subsequent to Crutzen's 2006 article, the Royal Society's 2009 *Geoengineering the Climate* report and the NRC 2015 *Climate Intervention* two-volume report have both marked significant advances in the attention to and perceived legitimacy of geoengineering as a concept. The very existence of these reports, *irrespective of content*, has contributed to a sense of increased legitimacy for the field of geoengineering as can be seen in journalistic accounts. Before the National Academy report was even published, its pending publication was referenced as evidence of geoengineering's move toward mainstream. Joel Achenbach writes in the *Washington Post* the week before the report's release: "That an institution as lofty as the National Academy of Sciences would take seriously an idea as dramatic as geoengineering is a sign of how little progress has been achieved in efforts to mitigate climate change" (Achenbach 2015). In earlier anticipation of the report's release, the *New York Times* article by Henry Fountain, previously quoted, indicates that there has been a clear and dramatic move from the fringe—("stuff of wild-eyed fantasies") to the mainstream ("now being discussed seriously by scientists") with the evidence of seriousness and legitimacy clenched simply with the fact that the National Academy of Sciences has taken on the issue for review (Fountain 2014).

An editorial examining geoengineering in the *Guardian* states:

> It's tempting too to dismiss ideas like pumping sulphate particles into the atmosphere or making clouds whiter as some sort of surrealist science fiction. But beyond the curiosity lies actions being countenanced and discussed by some of the world's leading scientific institutions. (Readfearn 2014)

This article includes a subsection entitled "Geoengineering on the table." As evidence that geoengineering is now being taken seriously, this section points to the geoengineering report by the Royal Society ("the world's oldest scientific institution"), the IPCC 2014 report that addresses geoengineering "in several chapters," and the National Academies of Sciences report (forthcoming at the time) addressing "'technical feasibility' of some geoengineering techniques" (Readfearn 2014). While this editorial takes a critical approach on geoengineering, the fact that these organizations are even considering it is used as evidence of the seriousness of the field. The inherent prestige and credibility of these institutions are taken prima facie to show that the mere consideration and discussion of the topic as a serious strategy among these organizations indicates that geoengineering has shifted to mainstream consideration and is now "on the table" (Readfearn 2014).

As discussed in the introduction, what climate solutions are even considered as options "on the table" has been a significant factor in climate politics to date. Steven Lukes (2005) theorized that power can be understood as three-dimensional: the power to achieve a desired outcome, the power to include or exclude an option from consideration, and the power to exclude an option even from consciousness. In terms of climate-change politics, the second dimension of power, "confining the scope of decision-making to only those issues that do not seriously challenge their subjective interests," has been employed by the American conservative movement to the effect of obstructing meaningful climate mitigation policy (McCright and Dunlap 2010, 106; see also Lukes 2005). Geoengineering being ascribed as now "on the table" is a new development in these climate politics. It is signaled by the scientific community in statements and especially through science policy reports as exemplified by the quoted selection above. At another level, the notion that geoengineering is now an option within the scope of consideration is also reinforced through public discourse as seen in the case of news media. Building on this conception that scientific reports signal geoengineering as "on the table" according to reporting, the following section will take a closer look at how news media directly report upon a key scientific report using the case study of the National Academies 2015 report.

MEDIA TREATMENT OF THE NATIONAL RESEARCH COUNCIL GEOENGINEERING REPORT

As discussed in the previous section, science policy reports allow experts in an emerging field to encapsulate current thinking within the field and express it to policy makers and the public. It is through news media, however, that the content within scientific reports primarily gets translated and repackaged for public consumption. The 2015 two-volume *Climate Intervention* report by the NRC of the National Academy of Sciences is the most substantial science policy report focused on geoengineering to date. This makes it an appropriate case study for examining in more depth how news media interact with science policy reports.

Within the corpus of 94 geoengineering news articles studied in detail here, 25 articles discuss the 2015 NRC *Climate Intervention* report. Of these, seven speak of the report in anticipation before its release, six are specifically focused on discussion of the report upon the release of the prepublication copy in February 2015, and another 12 articles reference or discuss the report following its release. The selection of articles discussing the NRC report are from the *New York Times*, the *Guardian*, the *Nation, Slate*, the *Washington Post, Scientific American, Ars Technica*, the *San Jose Mercury*, and *National Geographic*, which provide a microcosm encompassing many of the themes seen in journalistic reporting on geoengineering in general.

As discussed, the mere existence of these reports contributes to the idea of increased legitimacy of the field. Through direct and implicit language, the fact that such a respected institution as the National Academy of Sciences has undertaken study of geoengineering is used to indicate a move toward mainstream legitimacy of the field. On the other hand, the theme of controversy, which will be discussed in the next chapter, also emerges in the discussion of the report. Journalists tend to cite individuals with strong feelings on the reports, providing a window into some internal controversy. This section considers questions like how the report is characterized, what elements of the report are considered, which experts are most cited in articles discussing the NRC report, and how the primary recommendations given in the report are relayed by journalists and editorialists for consumption by the public.

To begin, how much of the NRC report is actually considered by authors when writing news articles in regard to it? It is generally understood that the summary is the most influential section of a policy report since many readers may rely heavily or exclusively on the summary. This certainly appears to be accurate in regard to journalists' discussion of the NRC *Climate Intervention* report. Within the selection of articles, the entirety of direct quotes and paraphrasing from the report can be attributed to the "Summary" or "Preface" sections. There are certainly practical reasons for why journalism in response to the report would limit specific references to the summary sections. First, the news articles that are the most focused on the report are those that were published the week of its release. One would not expect the writers to have had the opportunity to read the report volumes in their entirety, as the two volumes together exceed 346 pages of dense scientific and technical content. Second, the news articles that discuss the report later in 2015 or 2016 also rely upon the summary sections, but since these articles are not focused primarily on the NRC report, specific discussion of the report in these articles tends to be less in-depth. In either case, for practical purposes, it seems that it is from the "Preface" and "Summary" sections that reporters and editorialists draw to discuss the report's content. This fact underscores the argument in the previous section that the content and framing (what is said and how it is said) in the summary sections of science policy reports are particularly salient for how the content and takeaway points of the reports are considered and interpreted.

Moreover, the report-originated quotations tend to be sparse and short in most general-audience articles. Within the studied corpus, only six of the 18 articles that discuss the NRC report after its release include direct quotations from the report itself. Other sources relied upon to discuss the report contents include the National Academy press release and press briefing, members of the research committee, reviewers of the report, unaffiliated scientists, social scientists, and other journalists. Indicating a breadth of sources, 28 different individuals are cited within the news articles analyzed here that discuss the report following its release. Most of these people are cited in only one article each, meaning there is a diverse array of expertise and opinions relied upon to complement content drawn from the report.

Within the corpus of articles discussing the report, the most oft-quoted person is NRC committee member Raymond Pierrehumbert who is quoted in five separate articles. For comparison, Marcia McNutt, the chair of the Committee on Geoengineering Climate which was responsible for authoring the report, is quoted by three separate articles. Of the 28 individuals cited, nine were members of the committee responsible for the report, and seven of these were each quoted by only one news article in the selection. Individuals acknowledged in the NRC report as reviewers were also cited: Climatologist Alan Robock (three articles), David Keith (three articles), and Clive Hamilton (two articles). Unaffiliated science journalist Eli Kintisch and climate politics author Naomi Klein were each cited multiple times as well. This means there was a broad spectrum of possible opinions and interpretations to color discussion of the report.

To assess whether the citation trends within this qualitative study match broader patterns, the LexisNexis news search engine was again used to investigate the extent to which these and other individuals were cited by all English-language news articles within the database. There were numerous individuals cited in news articles that discuss the release of the report, but certain voices emerge as frequently evoked in news media while others remain absent or sparse. Table 4.3 shows the percentage of articles that reference or cite relevant individuals among geoengineering-related articles that discuss the NRC report the month of its release within the LexisNexis database. All of the individuals listed in table 4.3 were also cited within the corpus of articles analyzed in-depth qualitatively, however, with some variation in the listing order.

Table 4.3   Most Cited Individuals Within News Articles Discussing NRC Report in February 2015, the Month of Its Initial Release

Individual Cited	% Articles Referencing
Marcia McNutt**	33%
Alan Robock*	33%
Raymond Pierrehumbert**	22%
David Keith*	19%
Clive Hamilton*	19%
Ken Caldeira**	14%
Waleed Abdalati**	14%
Naomi Klein^	8%
Eli Kintisch^	6%

Note: Based on newspaper articles within the LexisNexis database of English-language news articles for February 2015 that include geoengineering terms plus "National Academy of Sciences" or "National Research Council," controlled for high similarity. There were 36 such articles. The values reported here are the percentage of these articles that cited to or referenced each individual. Search was run in 2018. ** = Committee member; * = Report Reviewer; ^ = Unaffiliated

Of the 16 members of the committee, eight members are cited by any newspapers in the database and only five of these members in multiple articles. Frequent reference to Marcia McNutt would be expected as she was chair of the committee responsible for the report. Other than McNutt, Raymond Pierrehumbert is the most cited member of the committee. He is a unique Committee member, being outspoken in his criticism of geoengineering and also providing a sharp commentary on the collaborative process of developing the report, to the point of questioning its coherence. An article in the *Washington Post* emphasized controversy regarding geoengineering and cited Pierrehumbert toward this end:

> "It will come as no surprise that there were very, very vigorous discussions by people on the committee who had very different viewpoints," said committee member and University of Chicago climate scientist Ray Pierrehumbert. "Once the report is out, it'll be a free-for-all in figuring out what the report actually means." (Achenbach 2015)

Similarly, in the Opinion pages of the *New York Times*, Andrew C. Revkin writes:

> I loved what the climate scientist Raymond Pierrehumbert had to say in *Slate* yesterday. His views are particularly notable not only because he was one of the report's authors but also because of his unbridled language in describing the process and his conclusions:
> "The nearly two years' worth of reading and animated discussions that went into this study have convinced me more than ever that the idea of 'fixing' the climate by hacking the Earth's reflection of sunlight is wildly, utterly, howlingly barking mad. In fact, though the report is couched in language more nuanced than what I myself would prefer, there is really nothing in it that is inconsistent with my earlier appraisals." (Revkin 2015)

As indicated here, Pierrehumbert proactively voiced his thoughts upon the publication of the report, providing candid and colorfully articulated concerns and critiques to be quoted by journalists. He uniquely embodied being a member of the committee that was responsible for the geoengineering report, while remaining consistent in his outspoken critique of geoengineering proposals.

Of the ten individuals acknowledged as reviewers of the report, four of them are cited within the LexisNexis English-language newspapers results. David Victor is cited once, while Alan Robock, David Keith, and Clive Hamilton are cited in multiple articles. David Keith is the most cited person in geoengineering, so it is not surprising that he is oft-cited in articles that also discuss geoengineering's definitive report. Within public discourse of geoengineering, Keith seems to ensure that his voice is frequently heard. Oft-cited by journalists, he seems to make himself available for comment, and he is also author of editorials (one of which will be discussed below) and a general-audience book advocating for geoengineering (*A Case for Climate Engineering*, 2013). Clive Hamilton has also been a consistent voice in geoengineering commentary and is author of an influential book on the subject, *Earthmasters*. Climate scientist Alan Robock has been a longtime critic of geoengineering and published in 2008 an influential article in the *Bulletin of the Atomic Scientists* entitled "20 Reasons Why Geoengineering May Be a Bad Idea" (Robock 2008).

Alan Robock took on a specific role in his correspondence with journalists with respect to the NRC report, acting as a sort of whistleblower in regard to the role of the U.S. Central Intelligence Agency (CIA) in partially funding the report and showing interest in details of albedo modification. This is a primary focus of the majority of articles citing Robock in regard to the NRC reports. An article in the *Guardian* revolves around this issue:

> Alan Robock, a climate scientist at Rutgers University in New Jersey, has called on secretive government agencies to be open about their interest in radical work that explores how to alter the world's climate. [. . .] "The CIA was a major funder of the National Academies report so that makes me really worried who is going to be in control," [Robock] said. [. . .] Robock said he became suspicious about the intelligence agencies' involvement in climate change science after receiving a call from two men who claimed to be CIA consultants three years ago. "They said: 'We are working for the CIA and we'd like to know if some other country was controlling our climate, would we be able to detect it?' I think they were also thinking in the back of their minds: 'If we wanted to control somebody else's climate could they detect it?'" (Sample 2015)

A number of news articles around the time of the NRC report release quote Robock in disclosing the role of the CIA in funding the report (at least 11 articles published in February 2015) and also details of the phone call and his concerns regarding CIA interest in geoengineering endeavors (at least five articles published in February 2015). Citing to Robock, like Pierrehumbert, advances characterization of controversy. These scientists provide voices of opposition and critique within the field to counter advocating voices, such as David Keith and others, so often included.

A number of unaffiliated individuals were asked to provide commentary in various news articles, some more than others. Eli Kintisch showed up in multiple instances in the corpus analyzed and in the database results. Kintisch has been a consistent voice providing commentary on the field of geoengineering, authoring one of the most influential general audience books on the subject, *Hack the Planet*, and providing nuanced critique of geoengineering proposals and possibilities. Karl Mathiesen writes in the *Guardian*: "science writer Eli Kintisch called geoengineering 'a bad idea whose time has come'" (Mathiesen 2015). Naomi Klein is evoked as geoengineering critic, having devoted a book chapter to the subject (*This Changes Everything*, 2014) and otherwise being outspoken in her critique. She is an example of a figure who is sometimes evoked merely as a personality representing critics, without necessarily being quoted directly. Extending the metaphor of voices of geoengineering, she becomes a face but not a voice in such articles.

Overall, the news reporting on the NRC report mostly accurately presents the committee's primary recommendations. Within the corpus, all six of the articles primarily concerned with the report at its time of publication, in addition to several subsequent articles, accurately relay the committee's emphasis on the importance of continued mitigation and adaptation efforts.[7] For example, Andrew Revkin writes in the *New York Times*: "The panels' overarching bottom line is straightforward: 'There is no substitute for dramatic reductions in the emissions of CO_2 and other greenhouse gases to mitigate the negative consequences of climate change, and concurrently to reduce ocean acidification'" (Revkin 2015). Craig Welch writes in *National Geographic*: "Committee members were blunt in their first recommendation: The world should focus first and foremost on curbing fossil fuel emissions rather than on any kind of geoengineering" (Welch 2015). Karl Mathiesen writes in the *Guardian*: "A report released on Tuesday by the US National Academies of Sciences (NAS) said tinkering with the global climate now would be 'irrational and irresponsible' and climate change can only be avoided by cutting emissions" (Mathiesen 2015).

All of the articles focused on the report's release also accurately conveyed the overarching recommendation for increased research into geoengineering, as did a number of others. Mathiesen goes on to say that, despite the caveats, "the influential group of 16 scientists who authored the report urged policy makers to commit to further research into some geoengineering techniques" (Mathiesen 2015). Lisa Krieger writes: "The council recommended a research agenda for how to offset our release of billions of tons of carbon dioxide a year caused by the burning of fossil fuels" (Krieger 2015). Dan Kahan writes: "Last week the National Academy of Sciences made headlines by calling for stepped-up research into geoengineering" (Kahan 2015).

The details and meaning of the research agenda vary between different articles' framings, but the majority that discussed the recommendations in detail recognized the committee's differentiation of CDR and albedo modification. Newspaper articles varied on whether they discussed the committee's positions regarding pursuit of CDR and/or albedo modification in detail. One or the other was commonly absent in later articles that were not specifically focused on the report.

There is room for ambiguity and differing interpretations regarding some of the report's recommendations, especially in regard to albedo modification. Nils Markusson, in his analysis of tensions in geoengineering reports, indicates there are tensions in regard to the framings of geoengineering within and between relevant reports (Markusson 2013, 4). Diversity between documents results from differing opinions and "attempts at persuasion to particular viewpoints," while ambivalence within documents "is caused by groups of authors trying to seek agreement on a text and express a coherent framing in spite of their differences [or as] the result of trying to pre-empt or entice responses from expected and imagined audiences" (Markusson 2013, 7).

Some of this ambivalence can be seen in regard to the NRC report's recommendations on albedo modification. The committee's "Recommendation 3" is that "Albedo modification at scales sufficient to alter climate should not be deployed at this time," while "Recommendation 4" is that "an albedo modification research program be developed and implemented that emphasizes multiple benefit research that also furthers basic understanding of the climate system and its human dimensions" (National Research Council 2015b, 148, 152). Within one *New York Times* article discussing the report, two scientists are quoted providing very different interpretations of the committee's recommendations regarding albedo modification:

David Keith, a researcher at Harvard University who reviewed the reports before they were released, said in an interview, "I think it's terrific that they made a stronger call than I expected for research, including field research." Along with other researchers, Dr. Keith has proposed a field experiment to test the effect of sulfate chemicals on atmospheric ozone. [. . .] Dr. Keith agreed, adding that he hoped the new reports would "break the logjam" and "give program managers the confidence they need to begin funding." [. . .]

Raymond Pierrehumbert, a geophysicist at the University of Chicago and a member of the panel, said in an interview that while he thought that a research program that allowed outdoor experiments was potentially dangerous, "the report allows for enough flexibility in the process to follow that it could be decided that we shouldn't have a program that goes beyond modeling." (Fountain 2015)

These two individuals quoted, of course, represent very different positions in regard to the appropriate trajectory for albedo modification. Pierrehumbert, in the quotation above, speaks to this idea of ambivalence, which he characterizes as "flexibility."

Nevertheless, most of the news articles picked up on the NRC's reluctance and very caveated position regarding albedo modification. All but one of the articles that discussed this issue in any detail pointed to concerns and caveats couching recommendations regarding albedo modification. David Biello characterizes the research recommendation as saying, "We should study up on climate interventions but focus the majority of efforts on thinning the blanket of CO_2" (Biello 2015). Scott Johnson says: "When the National Academy of Sciences report on geoengineering, released last week, looked at techniques to reflect some sunlight away from the Earth to counteract anthropogenic warming, the result wasn't exactly a glowing appraisal" (Johnson 2015). Andrew Revkin points to the report's consideration of "geoengineering prospects and concerns—the concerns mainly being about adding sun-blocking particles to the atmosphere" (Revkin 2015). Referring to the report summary, Joe Nocera states:

> The reports concluded that, while "climate intervention is no substitute for reductions in carbon dioxide emissions," the politics around carbon reduction have been so fractious that the day could well come when geoengineering was needed as part of a "portfolio" of responses to global warming. It urged further study for both methods, and, in particular, called for the establishment of a research program to examine the possible risks of solar radiation management. (Nocera 2015)

In regard to relaying the committee's caveated position on proceeding with albedo modification research, the one exception within the corpus studied was an editorial by none other than David Keith. While other editorialists as well as journalists acknowledged the reservations the committee signaled regarding albedo modification, Keith's framing is singular. His editorial entitled "Why We Should Research Solar Geoengineering Now" was not specifically focused on the NRC report, but it does engage with the report and claims that the NRC 2015 report, along with previous consideration by the National Academy, legitimizes and promotes field studies of albedo modification:

> Because the warming impact of carbon is more or less forever, all that we can achieve this century by cutting emissions is to stop making the problem worse. Solar geoengineering allows a more optimistic outcome. In combination with technologies to remove carbon that is already in the atmosphere, it would allow humanity to aim to restore the preindustrial climate over two human lifetimes.
>
> Despite this promise, there is little organized research on solar geoengineering. The U.S. National Academy of Sciences highlighted the potential of solar

geoengineering in 1982. It delved deeper in 1990 and again in January 2015, when it recommended a broad research program and suggested that small-scale outdoor experiments could yield valuable knowledge. (Keith 2016)

Here, Keith disproportionately emphasizes a recommendation of field experiments as a takeaway from the NRC report. In the report, the committee makes an unqualified recommendation to advance CDR research. However, its discussion around albedo modification research leaves open whether outdoor experimentation is recommended. As Pierrehumbert suggests, the reports leave room "that it could be decided that we shouldn't have a program that goes beyond modeling" (cited in Fountain 2015). Keith's editorial article lists only "small-scale outdoor experiments" for "solar geoengineering" in detailing the committee's recommendation for "a broad research program," implying a mandate on outdoor experimentation from the report, which is a contestable interpretation of the committee's research recommendations.

As a microcosm of broader geoengineering public discourse, news articles engaging with the NRC science policy report simultaneously portray various tensions. The consideration of geoengineering prospects by preeminent research academies is used to indicate increasing legitimation of the field. At the same time, however, a sense of controversy among scientists and others within the field is emphasized in news media. Which voices receive outlet in news media is part of this, with individuals representing strong opposing positions often invoked for citations and references. Moreover, the risks and, at times, outlandishness of geoengineering proposals are highlighted. Finally, interpretation and presentation are closely linked to how writers frame relevant content, as can be seen in how various authors present the highlights and recommendations of the NRC report.

In his study of high-profile geoengineering reports, Nils Markusson argues "that ambivalence, together with diversity, is key to the analysis of socio-technical imaginaries, and indicative of attempts at forging new relationships around the geoengineering imaginary" (Markusson 2013, 4). News media, in repackaging viewpoints on geoengineering for public consumption, provide an extra layer of potential ambivalence and diversity in terms of considering the geoengineering imaginary. Through their presentation, news media not only reinforce and legitimize certain geoengineering narratives but can also reshape them in the process. In grappling with some of the nuance and ambiguities involved in reporting on a nascent, evolving, and contested field that exists largely in theory as opposed to material manifestations, new strands of "ambivalence" or "diversity" emerge in regard to the socio-technical imagination of geoengineering.

CONCLUSION

Through analysis of a broad corpus of news media, this study contributes further depth of understanding to public discourse around geoengineering as encapsulated within the core artifact of written media treatment of the topic. News articles have both represented and articulated increased attention and interest in geoengineering, particularly in the 10-year period of 2006–2016. The move of geoengineering consideration from "fringe" to mainstream, coinciding with key documents within the field itself (particularly Crutzen's 2006 article, the Royal Society's 2009 report, and the NRC's 2015 report), is paralleled by the news coverage of it. As discussed above, mainstream journalism overall accurately presented the big take-aways from science reports. While there is convergence of discourse around geoengineering from both science policy reports and journalistic articles, there are also trends that are unique to the genre comprised of news reporting and editorials. The next chapter examines discursive conventions specific to news media's engagement with geoengineering in presenting it to the public.

NOTES

1. Articles from all dates through 2016 were considered for the qualitative analysis. Searches for relevant articles were made through the Google News search engine as well as through individual mainstream publications' archives. The resulting corpus of articles is inclusive and representative, but not exhaustive, of the universe of possible geoengineering articles in mainstream news reporting. The qualitative coding software, Nvivo, was used to demarcate and track relevant themes, subjects, actors, and linguistic or discursive practices.

2. A search was created to identify and analyze geoengineering-related articles in the broad array of English-language international newspapers included in the LexisNexis news database. Search terms, of which variations were tested and the list optimized to maximize relevant results and minimize false positive results, were: Climate OR warming AND geoengineer! OR geo-engineer! OR climate engineer! OR solar radiation management OR albedo enhancement OR albedo modification OR carbon dioxide removal AND NOT Senergy OR Seismic. Search was run in LexisNexis Academic in May 2017. The search by year was repeated in January 2021 with the result of slight variation, with some years off by a few articles, presumably due to minor changes in the database, but the trends and relative change over time were fully consistent. The purpose of this component of the research is primarily to show change over time, so the exact numbers are not the key takeaway, but rather repeating the search for each year for all English-language news articles discussing geoengineering in the LexisNexis database highlights relative changes over time.

3. These figures are based on searching for "geoengineering" on the individual publications' websites and then qualitatively assessing which articles include relevant discussion of geoengineering as the term relates to the present topic.

4. Data based on searching the LexisNexis database by year with a search list optimized to capture relevant articles while minimizing false positives. The figures are based on the resulting search results, refined to newspapers only and controlled for high similarity.

5. Keith and Caldeira are also among the expert witnesses called upon to testify before Congress on the topic of geoengineering, as will be discussed in the next chapter. Tangentially, it is an interesting twist upon their concept of the "geoclique" that both Eli Kintisch and Clive Hamilton themselves are now among some of the individuals most cited in geoengineering public discourse (see table 4.1).

6. Incidentally, there are only two women on the list of 15 frequently cited informants. Marcia McNutt in her capacity as committee chair is the most frequently cited woman, followed by Jane Long formerly of Lawrence Livermore National Laboratory, who receives references in 0.4% of the articles within the 10-year period. Moreover, both of the two women on the list, experts in their fields, receive significantly less citations than even Bjørn Lomborg, the contrarian political scientist who plays the skeptic on climate policy.

7. The committee's first recommendation was: "Efforts to address climate change should continue to focus most heavily on mitigating greenhouse gas emissions in combination with adapting to the impacts of climate change because these approaches do not present poorly defined and poorly quantified risks and are at a greater state of technological readiness" (National Research Council 2015, 3).

Chapter 5

News Media Framing and Discursive Presentation of Geoengineering

The previous chapter examined news media's engagement with geoengineering over time, including what voices dominated and what narratives prevailed. This chapter will continue the analysis of journalism's treatment of geoengineering by looking more closely at the discursive elements and specific substantive elements that together create the frame in which geoengineering is presented to the public.

Just like the framing of climate change itself has had demonstrable effects on public perceptions and policies (e.g., Oreskes and Conway 2010; Buell 2003; McCright and Dunlap 2010), the framing of technological proposals as possible responses to it can shape public discourse in terms of characterizing the pursuits, delineating the scope of available options, and differentiating the relative legitimacy of these possible responses. As Matti Luokkanen, Suvi Huttunen, and Mikael Hildén pointed out in their thoughtful study of geoengineering metaphors: "For a wider public audience these complex issues are new" so "terminology and conceptualization are important in influencing the basic understanding of the issue" (2014, 967).

The conceptualization, terminology, and metaphors used in presenting geoengineering to the public influence public perception, understanding, and assessment of these emerging technologies. Along with these discursive elements, substantive elements also frame the discourse and terms of deliberation. These include who is quoted for expert explanation and the scope of climate abatement or geoengineering proposals enumerated in describing options available. In selecting which scientific voices and narratives are brought from within the field and broadcasted to a public audience, news media contribute to reinforcing certain notions of legitimacy over others.

As discussed in the previous chapter, public discourse around geoengineering has included a narrative emphasizing a move toward mainstream from the realm of so-called fringe science in conjunction with a lifting of a taboo within mainstream science. The move toward mainstream, however, does not come without controversy. As will be discussed in the following section, the discursive theme of emphasizing controversy arises with the theme of geoengineering's mainstreaming. Once geoengineering is treated as a mainstream consideration within popular discourse, a space is opened up for portraying the competing notions of legitimacy between and among advocates and critics.

FOCUS ON CONTROVERSY

In contrast to science policy reports, popular media presentations on geoengineering often highlight controversy. This is not surprising, given that controversy provides for an engaging literary effect to interest readers. Furthermore, as customary in journalism, there is a tendency to present two sides of a story. For geoengineering, that means there is a presentation of controversy between proponents and critics both within and beyond the scientific community. The words "controversy" or "controversial" are consistently used to describe the field of geoengineering and specific geoengineering proposals in both American and British media, with 14% of the sampled articles explicitly using these terms in characterizing the field.

For example, Richard Black, environmental correspondent for BBC News, repeatedly emphasizes controversy in his articles focused on geoengineering in 2012. Black writes:

> Few issues arouse as much controversy in environmental circles these days as geoengineering "technical fixes" to tackle climate change, by sucking carbon dioxide from the air or by reducing the amount of sunlight hitting the Earth. (Black 2012b)

In a separate article, Black again emphasizes controversy pointedly:

> The field of implementing technical climate fixes, or geoengineering, is full of controversy, and even those involved in researching the issue see it as a last ditch option, a lot less desirable than constraining greenhouse gas emissions. [. . .] Adding to the controversy is that some of the techniques proposed could do more harm than good. (Black 2012a)

As indicated in this excerpt, controversy is inherent to geoengineering, in part due to the inherent risks of proposed technologies. While it might seem obvious that geoengineering is controversial due to its global scope and the level of risk involved, what is interesting is that within news media—a proxy for public discourse on the topic—the theme and narrative of controversy was not always a given.

As discussed in the previous chapter, one of the narratives common around geoengineering is its fringe origins and subsequent move toward mainstream. The emphasis on controversy within the sampled corpus of geoengineering-focused newspaper articles is clustered largely around 2012–2015, with all but one of the sampled articles (2010) explicitly referencing "controversy" published in this time frame. Thus, explicit highlighting of "controversy" arises within the sample after "fringe" has become a past-tense descriptor (the transition of which occurred in 2010 as discussed in the previous chapter), in a sense taking its place in the storyline of interpersonal intrigue. This suggests that controversy becomes a core component of the geoengineering narrative only after it becomes sufficiently mainstream to warrant contention between competing notions of legitimacy.

However, to better understand when the theme of "controversy" entered the public narrative in news reports on geoengineering, the LexisNexis search engine was again used to compare the analyzed corpus with a wider selection of English-language news.[1] Among articles discussing geoengineering in the LexisNexis database, from 1995 through 2005, none used the words "controversy" or "controversial." This changed with a dramatic spike to 26% of geoengineering-relevant articles in the database using one of these terms in 2006. From 2007 through 2016, 16% of articles with any discussion of geoengineering in the LexisNexis database use a form of the word "controversy," with some variation by year (see figure 5.1). This indicates that "controversy" became a common theme within public discourse on geoengineering starting in 2006, the year of Paul Crutzen's influential article, which has been ascribed

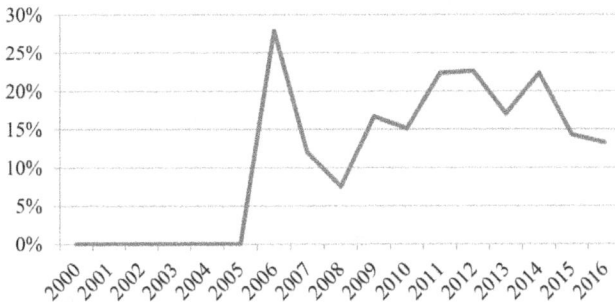

Figure 5.1 Percentage of News Articles Discussing Geoengineering That Mention Controversy, by Year (Based on English-Language News Articles Included in LexisNexis Database). *Source*: Author.

as breaking the "taboo" on discussing geoengineering and driving the concept into mainstream consideration. Moreover, the spike from 0% to 26% suggests the concept did not enter the discourse gradually but arose suddenly as a new discursive theme that became quickly established.

By either measure, "controversy" enters the narrative once the transition from "fringe" toward "mainstream" discourse begins. Some sampled articles even explicitly make the connection between scientific mainstreaming and controversy. For example, the quintessential article by Henry Fountain, climate journalist for the *New York Times*, states:

> Such ideas for countering climate change [. . .] are now being discussed seriously by scientists. [. . .] That does not mean that such measures, which are considered controversial across the political spectrum, are likely to be adopted anytime soon. (Fountain 2014)

Similarly, Lisa Krieger, science writer for the *Mercury News*, integrates both the themes of mainstreaming and environmental necessity with that of controversy:

> Scientists are so concerned about global warming that they're now calling for tests to find ways to cool the planet—the first step toward exploration of a highly controversial field that sounds like science fiction. (Krieger 2015)

In this case, the scientific mainstreaming is implied to have already occurred and that "scientists" (stated in a rhetorically general sense) are promoting the advancement of geoengineering due to environmental necessity.

Once controversy becomes part of the common narrative, it is taken for granted and spoken of in bold and absolute terms. Controversy is spoken of as an essential component of the technology: geoengineering "is" controversial (as opposed to being "considered" controversial or any other mitigating terminology). Furthermore, not only is this controversy essential, but extreme. For example, in BBC's Richard Black highlighted the extent of controversy, stating: "Few issues arouse as much controversy" as geoengineering (Black 2012b). An article in the *Economist* emphasizes: "To say that geo-engineering is controversial is an understatement" (2013). An earlier *Economist* article paralleled this nascent technology with other highly controversial emerging technologies: "Like genetic engineering was in the 1970s, the very idea of geoengineering is controversial" (*The Economist* staff 2010b).

Once controversy is established as a discursive theme, there are various ways to go about detailing it, particularly in terms of framing the sides of the debate. To say there are only two sides of the geoengineering debate would be a gross simplification. A better conceptualization would be a spectrum of viewpoints ranging from strong advocates of geoengineering technologies through strong critics. The middle range of the spectrum would include a broad variety of positions including CDR advocates opposed to albedo modification, research proponents reluctant about deployment, and those with ambiguous positions due to concern that the technology is dangerous paired with concern that it may nonetheless be necessary. Some journalists capture a sense of this nuance. For example, Henry Fountain articulates that, in comparison to CDR, "solar radiation management, is far more controversial" (Fountain 2015). However, within much of news media, a two-sided debate is often implied through referencing proponents and critics in a point-counterpoint style. Depending on the vantage point of the writer, and this is more pronounced in editorial-style articles than news-style articles, the "sides" of the debate might be characterized very differently.

In setting up the sides of the debate, an interesting word choice is used by some journalists to describe opponents of geoengineering. While the term "skeptics" relating to climate issues has a long-standing meaning, referring to those who dismiss or attempt to problematize science regarding climate change, within geoengineering discourse the term has been appropriated to refer to those concerned about the prospects of geoengineering. This usage means that geoengineering "skeptics" quite often include environmentalists who believe in, and are very concerned about, climate change. Contrastingly, as will be discussed further, some of the strongest advocates for geoengineering are erstwhile climate skeptics (see Hamilton 2013c).

PRESENTING GEOENGINEERING AS A MITIGATION ALTERNATIVE

Tangentially related to the emphasis on controversy, as well as the journalistic trend of presenting "two sides" of a debate, is the theme of mischaracterizing geoengineering as an alternative to mitigation. Despite frequent reiteration from the scientific community, including proponents, that geoengineering is perhaps a supplement or a stopgap but not an alternative to reducing emissions, this subtlety is often lost in the presentation of geoengineering through popular media.

Science policy reports as well as individual scientists advocating for advancement of geoengineering are explicit that geoengineering, particularly albedo modification, is not an alternative to emissions abatement. Rather, geoengineering is commonly presented by experts in three manners: (1) a supplement to mitigation, (2) a stopgap measure to buy time for sufficient emissions reductions to be instated and become effective, especially in light of latency issues, and/or (3) a Plan B in the case of climate emergency due to insufficient mitigation. The importance of emissions abatement is consistently and clearly emphasized.

For example, the National Academies report states at the outset: "There is no substitute for dramatic reductions in the emissions of CO_2 and other greenhouse gases to mitigate the negative consequences of climate change, and concurrently to reduce ocean acidification" (National Research Council 2015a, 2). The Committee's "Recommendation 1" is:

> Efforts to address climate change should continue to focus most heavily on mitigating greenhouse gas emissions in combination with adapting to the impacts of climate change because these approaches do not present poorly defined and poorly quantified risks and are at a greater state of technological readiness. (National Research Council 2015a, 3)

In the context of albedo modification, it is emphasized:

> The less CO_2 that humans release to the atmosphere, the lower the environmental risk from the associated climate change and the lower the risk from any albedo modification that might be deployed as part of the strategy for addressing climate change. It is widely recognized that the possibility of intervening in climate by albedo modification does not reduce the importance of efforts to reduce CO_2 emissions. (National Research Council 2015b, 36)

Elsewhere,

> albedo modification is no substitute for mitigation. Hence, in order to avoid serious longer-term problems, any future decision to embark on aerosol injection should be paired with efforts to mitigate greenhouse gas emissions, remove carbon dioxide from the atmosphere, or both. (National Research Council 2015b, 145)

Similarly, the Royal Society report explicates: "Geoengineering methods are not a substitute for climate change mitigation, and should only be considered as part of a wider package of options for addressing climate change" (Royal Society 2009, 58). Both of these key geoengineering science policy reports clearly point to geoengineering as a potential supplement to mitigation and not a replacement or alternative to mitigation.

Among scientist advocates, the message tends to be the same. David Keith is one of the most vociferous and oft-quoted proponents of geoengineering from within the expert scientific community (see tables 4.1 and 4.2). He unabashedly advocates for consideration of solar geoengineering and not just as a Plan B like many frame it but rather he asserts: "Early use of solar geoengineering" should be considered to save certain ecosystems and slow carbon cycle feedbacks (Keith 2016). Despite being a strong proponent of albedo modification, Keith clearly states that "it is not a substitute for cutting emissions—it is a supplement" (Keith 2016). What he advocates is "a combination of cutting emissions and solar geoengineering" to stave off dramatic effects of climate change (Keith 2016).

Other scientist proponents of albedo modification make similar caveats. For example, space scientist Russell Bewick, who has researched possible strategies of space-based geoengineering, is quoted in a *LiveScience* article as stating: "I would like to make it clear that I would never suggest geoengineering in place of reducing our carbon emissions" (Choi 2012). Bewick goes on to specify his take on the temporal dynamic in regard to a particular space-based albedo modification scheme:

We can buy time to find a lasting solution to combat Earth's climate change. The dust cloud is not a permanent cure, but it could offset the effects of climate change for a given time to allow slow-acting measures like carbon capture to take effect. (in Choi 2012)

Advocates from other disciplines also specify their position that geoengineering is not a substitute for addressing emissions. Oliver Morton, a journalist who has extensively covered geoengineering and written a book on the subject, argues on its behalf as a serious climate solution, but he also clarifies, as quoted here: "I do not in any way see geoengineering as an alternative to a program of emissions reduction" (Morton 2015). Martin Bunzl, a philosopher engaged with geoengineering, also frames it as a strategy that would be taken in conjunction with emissions abatement: "We have to decarbonize. We can decarbonize with the option to geoengineer, or we can decarbonize without the option" (in Fischer 2009).

In these ways, Keith, Bewick, Morton, and Bunzl represent a particular strand of temporal treatment regarding the consideration of geoengineering: "early use" to prevent certain ecological effects of climate change (Keith 2016) and to "buy time" (Choi 2012; Fischer 2009; cf. Morton 2015) to "decarbonize the economy" (Fischer 2009). Another common narrative strand treats solar geoengineering as a Plan B on reserve "in case needed." Geoengineering researcher Hugh Hunt's perspective fits with this emergency scenario strand of the narrative: "You'd only consider doing it if it was a real emergency and there was no other solution. [. . .] But I do worry we're getting close to that situation" (in Pappas 2013). Both of these temporally disparate arguments in favor of pursuing albedo modification have in common the caveat that geoengineering does not replace the need for mitigation.

One of the reasons that discussion of geoengineering is often, and has historically, been minimized in climate discourse is the same reason that experts on the topic so frequently reemphasize that geoengineering would constitute a supplement rather than an alternative to mitigation: concerns of moral hazard. As mentioned in section I, moral hazard is "a term derived from insurance, and arises where a newly-insured party is more inclined to undertake risky behaviour than previously because compensation is available" (Royal Society 2009, 37). In the case of geoengineering, the concern is that a strategy that seems to present an alternative to mitigation would reduce the tenuous political and social motivation to address the emissions which are the cause of climate change. However, beyond concern of minimizing this risk of moral hazard, as presented in the range of examples above, among mainstream experts on geoengineering there is by and large consensus that geoengineering, especially albedo modification, independent of emissions abatement, would not be advisable or desirable even if feasible. In the words of the NRC report: "The Committee considers it to be irrational and irresponsible to implement sustained albedo modification without also pursuing emissions mitigation, carbon dioxide removal, or both" (National Research Council 2015b, 147).

Despite the insistence of scientists with expertise in the field that geoengineering would be a supplement, a stopgap, or a Plan B to mitigation, writers addressing geoengineering in popular media at times miss this point and can contribute to the false understanding that geoengineering may present an alternative to emissions abatement. For example, an article in the *Economist* states: "Geoengineering is an umbrella term for large-scale actions intended to combat the climate-changing effects of greenhouse-gas emissions without actually curbing those emissions" (*The Economist* staff 2010b). While this definition of geoengineering does not necessarily preclude the possibility that emission abatement may proceed simultaneously, it implies that geoengineering may constitute an alternative to "actually curbing" emissions. The article goes on to set up a dichotomy between those who favor mitigation and those who favor geoengineering: "Most of those who fear climate change would prefer to stop it by reducing greenhouse-gas emissions. Geoengineers argue that this may prove insufficient and that ways of tinkering directly with the atmosphere and the oceans need to be studied" (*The Economist* staff 2010b). By constructing a dichotomy between traditional mitigation-seekers and "geoengineers," there is a false implication that those the article refers to as "geoengineers" are not also strongly in favor of mitigation. As discussed above, even strong proponents of geoengineering solutions usually see them as a supplement, stopgap, or Plan B to emissions abatement.

A *Newsweek* article in response to the failed Copenhagen Climate Summit presents geoengineering as an alternative to addressing emissions:

There will be no climate treaty to emerge from the conference in Copenhagen this month, global leaders now concede. But there may be alternative ways to help combat global warming. Various methods of geo-engineering employ unorthodox means to cool the planet. Advocates say that some of these pro-posals could be implemented quickly and cheaply. One concept is known as stratospheric aerosol insertion. (Ellison 2009)

In this selection, the phrasing "to help combat global warming" leaves room for some ambiguity in the degree to which geoengineering stands alone or not. While this ambiguity is left to stand in the article, what is more striking in the framing is the construction of a juxtaposition between geoengineering and mitigation. The author explicitly uses the word "alternative" in relation to geoengineering set in juxtaposition to a climate treaty, which is implicitly tantamount to saying emissions mitigation.

Treating geoengineering as an alternative to mitigation (in the traditional sense of emissions reductions) is often a subtle or indirect implication as opposed to an explicit contention. For example, a *Newsweek* article demon-strates the tendency to imply that geoengineering is an alternative to mitiga-tion through comparison:

It sounded like a panacea for climate change: "geo-engineering" the atmosphere to block some sunlight and counter global warming [. . .] a quick fix to stabi-lize or even reverse the heating of the planet. It would head off worsening heat waves, droughts, and rising sea levels. The estimated price is right, too. A 2009 analysis found that geo-engineering would cost only $2 billion or so a year, chump change compared with converting from CO_2-producing coal, oil, and natural gas to wind, solar, nuclear, and biofuels. (Begley 2011)

By characterizing geoengineering as a "panacea" and a "quick fix," the author implicitly juxtaposes it against the arduous efforts involved in reducing emis-sions. Moreover, by comparing the price of albedo modification strategies to the cost of transforming energy systems, the author implicates that the two would be competing solutions as opposed to coexisting or supplemental as scientists tend to present albedo modification.

Yet another *Newsweek* article similarly implies albedo modification as an alternative to mitigation through cost comparison. Fred Guterl states: "A judicious application of sulfur dioxide to the upper atmosphere [. . .] would have an almost immediate impact on temperature. And it would cost a thousand times less than even the most optimistic scenarios for cutting emis-sions" (Guterl 2009). Particularly with albedo modification, for which cost is relatively low compared to mitigation, it is common for journalists and editorialists to indicate it would be a cheaper alternative. This provides the audience with the false sense of competing options.

Near the end of this article, Guterl discusses the political risk of moral hazard, again based on the evaluative factor of differential costs: "Success in lowering temperatures—or even the knowledge that scientists had the means to do so—might decrease the political will to make costly emissions cuts" (Guterl 2009). This moral hazard framing again juxtaposes costs as if there were an economic calculation to be made. Moreover, it implies the potential of geoengineering standing in the place of emissions reductions or threatening emission cuts through its very consideration. However, it should be noted that in this discussion, Guterl (2009) does ultimately acknowledge that "not even the most zealous advocate of geo-engineering argues for using it in lieu of cutting and capturing carbon."

Certainly, however, there are some exceptions in terms of parties who do consciously or intentionally present geoengineering as an alternative to emission abatement. This is primarily true of members or representatives of vested interests like the oil and gas industry or conservative think tanks with fringe positions. For example, Bjørn Lomborg, a political scientist and the self-titled "Skeptical Environmentalist" who thrives on controversy, as mentioned previously, has inserted himself in geoengineering discourse with arguments that geoengineering may be a better solution than mitigation. Lomborg represents the community of climate-skeptics turned geoengineering advocates. As Clive Hamilton wrote in the *New York Times*,

> Engineering the climate is intuitively appealing to a powerful strand of Western technological thought that sees no ethical or other obstacle to total domination of nature. And that is why some conservative think tanks that have for years denied or downplayed the science of climate change suddenly support geoengineering, the solution to a problem they once said did not exist. (Hamilton 2013c)

Tina Sikka's analysis of conservative think tank discourse discusses how advocates from these communities "construct geoengineering research and practice as necessary, commonsensical and natural" (Sikka 2012, 166). This community stands apart from the mainstream scientific community concerned with climate, including geoengineering advocates within it.

Within opinion pieces and editorials, some of the most forceful geoengineering advocates, like social scientist David Victor, indicate a viewpoint that geoengineering may be an alternative to emissions abatement despite rhetorical acknowledgment of the mainstream emphasis on emission controls. In an influential *Foreign Affairs* article promoting geoengineering, David Victor, along with M. Granger Morgan, Jay Apt, John Steinbruner, and Katharine Ricke, state that while geoengineering "cost estimates" tend to be "preliminary and unreliable [. . .], there is general agreement that the strategies are cheap" and could cost "as little as a few billion dollars, just one percent (or less) of the cost of dramatically cutting emissions" (2009, 69). Comparing the cost of albedo modification proposals to the cost of "dramatically cutting emissions" implies an either-or relationship between those two strategies for addressing climate change as opposed to a supplementary or complementary relationship.

These authors, who are among the strongest advocates of geoengineering, close the article by stating:

> The best and safest strategy for reversing climate change is to halt this buildup of greenhouse gases, but this solution will take time, and it involves myriad practical and political difficulties. Meanwhile, the dangers are mounting. In a few decades, the option of geoengineering could look less ugly for some countries than unchecked changes in the climate. Nor is it impossible that later in the century the planet will experience a climatic disaster that puts ecosystems and human prosperity at risk. It is time to take geoengineering out of the closet—to better control the risk of unilateral action and also to know the costs and consequences of its use so that the nations of the world can collectively decide whether to raise the shield if they think the planet needs it. (Victor et al. 2009, 76)

The statement that "the best and safest strategy for reversing climate change is to halt this buildup of greenhouse gases" at first seems consistent with the scientific consensus emphasizing the importance of emissions abatement. Likewise, the next phrase of the sentence regarding the emissions solution taking time appears consistent with the common narrative among geoengineering proponents within the scientific community who argue that albedo modification may be useful to pursue in conjunction with emissions abatement as a stopgap measure to buy time while emissions are dramatically reduced. However, the final phrase of this sentence, that "it involves myriad practical and political difficulties," paired with the subsequent sentences reorients the message. The authors imply that the "myriad of practical and political difficulties" challenging emissions mitigation efforts would not be mirrored in the political challenges also inherent in the pursuit of albedo modification. They fail to acknowledge that replacing this political-consensus defying strategy with geoengineering may repeat the same international and domestic challenges in a new realm.

On the other side of the spectrum, even conscientious science journalists who recognize that geoengineering is not a substitute to abatement make statements at times that imply otherwise. For example, in the opening paragraph of climate correspondent Karl Mathiesen's (2015) article in the

Guardian, entitled "Is Geoengineering a Bad Idea?" he critically states: "It is considered by many to be the ultimate admission of our failure to curb carbon emissions—a tech-fix that excuses continued carbon gluttony in the industrialised world." While elsewhere he notes the NRC position against geoengineering as a stand-alone strategy, he implies in the characterization of "a tech-fix that excuses continued carbon gluttony" that geoengineering is presented as an alternative to mitigation.

In presentation to the public by news media through both editorialist and journalistic accounts, a sense of ambiguity is relayed in regard to the relationship between geoengineering and emissions abatement. Within the corpus of articles focused on geoengineering, there is some recognition of the caveats raised by scientists that albedo modification would not replace the need for abatement. However, this point is often obscured by discursive moves that imply an either-or relationship, such as juxtaposition of geoengineering and mitigation as distinct choices or through the comparison of costs/effort that indicate geoengineering as a cheaper/easier option thereby implying the two would not necessarily coexist. The treatment of geoengineering within popular media as the other side of mitigation in a dichotomous relationship may be an extension of the journalistic tendency to portray two sides of an issue. The resulting tensions contribute toward reshaping the public narratives around the geoengineering socio-technical imaginary of what is possible, acceptable, and desirable.

THE DECOY EFFECT IN PRESENTING
GEOENGINEERING "OPTIONS"

One discursive theme particularly common in popular media covering geoengineering can be characterized as a decoy effect in presenting geoengineering options. This engaging rhetorical technique presents a range of options, including more extreme, controversial, or absurd proposals in addition to discussion of options being more seriously considered. There are two primary effects of the use of decoy options, the first simply being literary intrigue to engage readers on the topic of geoengineering through inclusion of more extreme, shocking, or science-fiction–like proposals. The second, and more important, is the potential to influence readers in their perspectives on geoengineering by making novel proposals seem less extreme by comparison.

What I call *decoy options* are geoengineering proposals not being seriously advocated but rather easily dismissed in favor of more popular proposals actually under consideration. By inclusion of more extreme and easily contestable decoy options in the presentation of geoengineering schemes, other options can be framed as more reasonable or benign by comparison. Contrary to the assumptions of many utility theories, which assume that "irrelevant alternatives" (or decoy choices) do not affect the outcome of rational deci-sion-making, decoys can have a significant effect in choice selection (Soltani, De Martino, and Camerer 2012). Hence, the common use of decoy options in geoengineering literature can contribute to normalizing, legitimizing, or otherwise reorienting the audience's thinking about particular geoengineer-ing proposals, whether incidentally or intentionally on the part of the authors.

Within the journalistic genre, the use of decoy or straw man options in discussion of geoengineering often appears as a literary technique to build intrigue in the story. Among geoengineering literature written for a general audience, there is a tendency to choose particularly extreme options, empha-sizing their enormity or even absurdity. For instance, in one article, the edi-tors of *Scientific American* introduce the range of geoengineering options by stating: "Scientists and engineers have proposed various approaches besides iron fertilization, such as hazing the skies with sulfates to mimic the cooling effects of a volcanic eruption or even launching a fleet of mirrors to deflect sunlight away from the planet" (Board of Editors 2015). The language of the last option, prefixed with the word "even" and with the enormity of the undertaking characterized by the description of "a fleet," makes clear that this option is not necessarily being presented as a serious consideration, but rather a decoy option that has the effect of increasing the scope of presented options, potentially making other options come across as more tenable than they would alone.

Another geoengineering article in the *Scientific American* presents the range of options in this way:

Some ideas are the stuff of science fiction: 15 trillion mirrors positioned in orbit to shield the planet from the sun's rays; a fleet of blimps 20 kilometers up feed-ing a constant stream of sulfur into the stratosphere; a navy of robot-controlled ships prowling the world's oceans, spraying seawater skyward to generate reflective clouds.

Others are more mundane: Plant trees to soak up carbon dioxide or paint roofs white to reflect sunlight. Most are unproven. All have major drawbacks. None offset ocean acidification. (Fischer 2009)

This presentation provides the two extremes of the range of options: those up-played in regard to their extremity or absurdity and those presented as "mun-dane," which include the least novel and lowest-risk options. Framing the range of options with such a widespread spectrum of risk and novelty, from the "stuff of science fiction" through the "mundane," has the discursive effect of acknowledging and discrediting the reader's potential starting assump-tions regarding the absurdity of geoengineering at large. It is emphasized that while some schemes are on the absurd end, the spectrum contains a wide array of options, including relatively lower risk and lower novelty proposals. Compared to "a navy of robot-controlled ships prowling the world's oceans," painting roofs white seems particularly tame.

The inclusion of decoy or straw man options is prevalent throughout articles in general audience publications, representing a wide range of positions (both critical and supportive of geoengineering) and publishers (from *Slate* and the *New York Times* to *Scientific American* and *Foreign Affairs*). The pervasiveness of the decoy effect is particularly highlighted by the disproportionate presentation of space mirrors as a potential albedo modification option. The geoengineering proposal of space mirrors is generally dismissed from serious consideration due to prohibitive costs. Moreover, there is no significant advocacy within the field for pursuit of this option. Yet, 25 of the 94 geoengineering-focused news articles include space mirrors or space reflection as one of the explicit geoengineering options presented. Incidentally, 25 articles also present MCB as a potential option. Notably, MCB is consistently considered among experts and proponents to be among the two most viable albedo modification strategies (National Research Council 2015b, 39, 113; Bellamy et al. 2012, 602), while space reflection is consistently dismissed as overly expensive and impractical for serious consideration (National Research Council 2015b, 104; Bellamy et al. 2012, 602; Royal Society 2009, 32–33). Yet, within the corpus, an equal number of articles list these two options in presenting the range of geoengineering options available for consideration.

Space reflection is a popular scheme for reference, particularly in general audience literature, precisely because it is considered intrinsically extreme or absurd and, as such, intriguing. Indeed, the idea of space mirrors is recurrently linked to an explicit comparison between geoengineering and science fiction. Exemplifying this trend of explicit comparison, Clive Hamilton writes in the *New York Times* that "some proposals, like launching a cloud of mirrors into space to deflect some of the sun's heat, sound like science fiction" (Hamilton 2013c). Elsewhere, Hamilton is quoted describing geoengineering schemes that "range from the benign [. . .] through to the science fiction, like putting a cloud of mirrors in space to deflect some sunlight" (Hamilton 2013b). Douglass Fischer writes in *Scientific American:* "Some ideas are the stuff of science fiction: 15 trillion mirrors positioned in orbit" (Fischer 2009).

In other instances, the extremity or absurdity of the space mirrors proposal is highlighted by particular wording such as the word "even" as in "proposals *even* include . . ." (emphasis is mine) This is seen in the *Scientific American* editorial list of geoengineering options ending with "even launching a fleet of mirrors to deflect sunlight away from the planet" (Board of Editors 2015) and in a *Slate* editorial critical of albedo modification which states, "Solar geoengineering seeks to reduce the amount of sunlight that warms the Earth at the surface, troposphere, upper atmosphere, or even space level" (Jospe 2016). One editorial presents examples of geoengineering options as follows:

> Geoengineering comprises technologies designed to counteract human-caused climate change: towering "carbon scrubbers" that would suck carbon dioxide from the atmosphere; the injection of iron pellets into the ocean to stimulate growth of carbon consuming phytoplankton blooms; or—my personal favorite—deploying zillions of mirror-coated nanotechnology flying saucers to form a stratospheric solar reflector. (Kahan 2015)

The absurdity of the solar reflection option is up-played with the language emphasizing the technological novelty ("nanotechnology flying saucers") and enormity of the effort (using the word "zillions" rather than a specific number range). While presented as absurd, this option is made to stand out, with the first-person aside "—my personal favorite—" used to draw particular attention to it.

One article, while itself subject to the trend, reflectively identifies a reason for the disproportionate representation of space mirrors among the range of options: "From a technological standpoint, the flashiest geoengineering scheme is space-based" (Pappas 2013). After then describing the basic technical process of the space reflection concept, the article goes on to say, "However, space-based schemes are the least likely to be implemented" due to their vast impracticalities as identified by experts (Pappas 2013). The emphasis on space-based schemes is explicitly recognized due to their being technologically flashy even though they are then dismissed as infeasible within reasonable time periods. This article, emblematic of two primary trends within the genre, focuses on presenting a range of geoengineering options, making them seem engaging and interesting, while also emphasizing controversy. It quotes primary scientists in each option area presented. Ultimately it elevates the options that are most commonly advanced among the range of options. Including the more extreme and "flashiest" options like space reflection frames the subsequent presentation of "less dramatic, and more feasible" options such as stratospheric aerosols and MCB, contextualizing their challenges as relatively surmountable (Pappas 2013).

Science journalism has conventionalized the decoy effect in part due to relaying the discourse of proponents and advocates who strategically employ the technique. For proponents of particular geoengineering proposals, the presentation of a range of options broadened by inclusion of more extreme or controversial ideas makes the favored options seem more reasonable by comparison. Embodied within public audience literature, journalists present the sense of homing in on options most favored by advocates within the geoengineering community. This is demonstrated in the *Scientific American* article in which the extreme "stuff of science fiction" is juxtaposed to the "more mundane," and ultimately the discussion is funneled down to focus on "the most favored option today [which] is the injection of sunlight-reflecting sulfur particles high into the atmosphere" (Fischer 2009).

Often journalists explicitly convey the preferred proposals of geoengineering advocates. In a review of the *Planet Remade*, science journalist Oliver Morton's book advocating geoengineering, another science journalist Thomas Sumner writes in *Science News Magazine*:

> The book lays out the typical laundry list of geoengineering proposals, from extracting carbon dioxide from the air to deploying giant Earth-orbiting space mirrors. But Morton has a clear favorite. A variety of airborne particles reflect sunlight like tiny disco balls. A fleet of high-flying planes could spray these aerosols into the stratosphere and thicken the sun-dimming veil that surrounds Earth. (Sumner 2015)

Similarly, in describing a speech on geoengineering by Lord Rees, Alok Jha writes:

> Geoengineering involves deliberate planet-scale interventions to counteract global warming. Techniques suggested include placing mirrors in space that reflect sunlight away from the Earth and fertilising the oceans with iron to encourage the growth of algae that can soak up atmospheric carbon dioxide. Other options include Rees's preference—to seed clouds in the upper layer of the Earth's atmosphere to bounce some of the sun's energy back into space. (Jha 2013)

These examples demonstrate the flow from decoy options to presenting the options being explicitly advocated by proponents. Both, of course, include space mirrors.

A more subtle version of this trend of homing in on a preferred option can be seen in the influential *Foreign Affairs* article advocating for more serious consideration and pursuit of albedo modification, written by geoengineering proponents David G. Victor, M. Granger Morgan, Jay Apt, John Steinbruner, and Katharine Ricke. They introduce the range of solar geoengineering options, opening with that of "putting reflective particles into the upper atmosphere, much as volcanoes do already," arguing this method would "offer quick impacts with relatively little effort" (Victor et al. 2009, 68). Their list of options continues with other proposals including "seeding bright reflective clouds" and "converting dark places that absorb lots of sunlight to lighter shades" noting that "engineered plants might be designed for the task" (Victor et al. 2009, 68). They conclude the list by stating: "More ambitious projects could include launching a huge cloud of thin refracting discs into a special space orbit that parks the discs between the sun and the earth" (Victor et al. 2009, 68).

In this way, their range of options begins with the favored proposal, stratospheric aerosols, and then goes on to list a number of other options of various levels of (in)feasibility or practicality, closing with space mirrors at the extreme end of the spectrum. These authors, who are proponents of albedo modification pursuit, positively frame the range of options presented. For instance, they use the positive term "ambitious" to characterize the space reflection category of albedo modification as contrasted to the many authors who emphasize its absurdity or novelty rather than its ambitiousness. While they do not explicitly dismiss any of the albedo modification options presented, they use the range of options to home in on their favored method, making clear that the alternatives are less feasible and practical than their preferred option. Immediately after this list, they emphasize the point by returning to the relative advantages of stratospheric aerosols, specifically stating that this approach "seems to be the easiest and most cost-effective option" and then continuing with a more detailed discussion of the topic (Victor et al. 2009, 69). This is a prime example of strategically presenting a range of options, including decoys, to suggest a sense of internal evaluation between options that ultimately advances the favored choice being promoted.

Even for authors more cautious or concerned about the prospect of geoengineering, the discursive trend of presenting a range of options broadened by decoys may have the effect of normalizing other options. For instance, Clive Hamilton, an Australian public intellectual who has written extensively on geoengineering, employs the decoy effect despite his tendency to lean toward caution and concern regarding geoengineering technology. In a *New York Times* Op-Ed, Hamilton writes: "While some proposals, like launching a cloud of mirrors into space to deflect some of the sun's heat, sound like science fiction, the more serious schemes require no insurmountable technical feats" (2013c). This is a quintessential presentation of a decoy option, emphasizing the absurdity of space mirrors through use of the science-fiction analogy, prior to opening the discussion of "more serious schemes" and their advantages. Yet, the article is not representative of a geoengineering proponent position but rather a concerned perspective in which Hamilton emphasizes the risks and uncertainties of geoengineering along with the potential mismatch of the "solution" to the cause of the climate crisis.

Furthermore, in his book on geoengineering, Hamilton includes an even more extreme decoy option than space reflection in his presentation of the range of geoengineering proposals: the "novel scheme to counter global warming" published in "the esteemed journal *Climatic Change*" that suggested "the effects of global warming could be countered by increasing the radius of the Earth's orbit around the Sun" (Hamilton 2013a, 3; referencing Jain 1993). These examples point to how the use of decoy options has become entrenched into geoengineering literature to the point where this discursive tool, which proponents may employ toward normalizing their favored geoengineering approaches, is even conventionalized and commonplace among authors with a neutral or critical stance on geoengineering.

ANALOGIES AND METAPHORS

Metaphors and analogies affect the interpretation, shaping, and reconstruction of the socio-technical imaginaries around technological responses to climate change. George Lakoff and Mark Johnson argue broadly in *Metaphors We Live By* that "most of our normal conceptual system is metaphorically structured; that is, most concepts are partially understood in terms of other concepts" (Lakoff and Johnson 1980, 56). Metaphors are particularly important for shaping conceptualizations of emerging technologies, which are subject to "interpretive flexibility"; as such, metaphors can "play an important role in the general framing of geoengineering" (Luokkanen, Huttunen and Hildén 2014, 978). Within scholarship analyzing geoengineering discourse, there are a few studies to date focused on metaphors within specific scope areas.

Nerlich and Jaspal (2012) examined metaphors within trade literature from 1988 to 2010. In the case of the industry trade literature genre, metaphors primarily served the purpose of promoting or positively framing geoengineering. The three "conceptual master metaphors they identify" are the planet as a body, the planet as a machine, and the planet as a patient/addict (Nerlich and Jaspal 2012, 131, 135). In their study, the metaphors used in geoengineering discourse were found to promote geoengineering and to frame geoengineering as a necessary option to have available (e.g., within the metaphor of earth as patient, geoengineering is likened to chemotherapy, citing David Keith to say no one wants it, but we want it available if needed) as well as feasible and doable (e.g., within the metaphor of earth as machine, comparing geoengineering proposals to how one would fix a car gives a sense of "easy" or "routine" feasibility) (Nerlich and Jaspal 2012, 137–139). Like other authors (e.g., Bellamy et al. 2013), they found that

> geoengineering metaphors and arguments [. . .] seem to be closing down debates about geoengineering and, in the process, debates about climate change mitigation, rather than opening them up. [. . .] This rhetoric limits social and ethical reflection on the issue of geoengineering by implicitly establishing the boundaries of "legitimate" debate. (Nerlich and Jaspal 2012, 142–143)

Luokkanen, Huttunen, and Hildén (2014) analyzed metaphors used in relation to geoengineering as presented through the *New York Times* and the *Guardian* from 2006 to 2011. In general audience news media, metaphors can be used in support or opposition to advancing the technology (Luokkanen, Huttunen and Hildén 2014, 978). The main metaphors identified within their corpus of study are (1) war and fight ("acting on climate change is like fighting a war," which is "commonly used in describing geoengineering neutrally"); (2) controllability ("geoengineering is like preparing for the future with insurance," with metaphors of controllability "mainly used to support further studies of geoengineering"); (3) mechanisms ("earth is like a machine and interventions on earth are like interventions in a machine's mechanism," which is often used in arguments against geoengineering); (4) health and illness ("the earth is like a living organism and geoengineering actions are like medical actions," which is often used in arguments against geoengineering) (Luokkanen, Huttunen and Hildén 2014, 973–977).

The corpus of articles reviewed in this study complements and builds upon these previous analyses of metaphor in relation to geoengineering with a broader scope of publications and time range. Because geoengineering is based upon hypothetical scenarios and therefore abstract to a general audience, metaphors and analogies are potent tools for explanation and framing. The present study affirms and contributes additional insights to the findings by Nerlich and Jaspal (2012) and Luokkanen et al. (2014) regarding the recurrent usage of mechanical and medical analogies.

A frequent analogy within surveyed news articles compares implementing albedo modification to setting a global "thermostat." Nerlich and Jaspal's study of trade literature considered the mechanistic metaphor as supporting arguments for pursuing geoengineering through presenting the earth as a machine that can be "fixed or repaired" (2012, 137). In contrast, Luokkanen et al. found mechanistic metaphors in newspaper articles to often be used in arguments against geoengineering (Luokkanen, Huttunen and Hildén 2014, 975). The present research on news media is consistent with this latter finding. In particular, within the corpus studied here, the common thermostat analogy is particularly used to allude to the potential of international conflict.

Over half (10 of 17 or 59%) of the articles studied here that employed the thermostat metaphor used it to raise the question of "who gets to set the thermostat" in the case that albedo modification is deployed, emphasizing the problematic global political challenges to agreeing upon a course for geoengineering. For example, Caroline Jones employs the thermostat analogy in an article entitled "The New Cold War: The Political Problem of Geoengineering," directly asking: "Which country's hand gets to rest on the global thermostat?" (Jones 2016). Speaking about the high barrier of challenges to instituting geoengineering, *New York Times* environmental journalist Andrew Revkin writes that among the challenges, "The main one is diplomatic, not technological. Who sets the thermostat?" (Revkin 2015). Emphasizing the potential for international conflict, one journalist writes: "There may be disputes over the 'right' temperature, setting off what's been dubbed 'the Thermostat Wars'—if Indonesia wants cooling to avoid sea level rise and Russia wants warming to increase agricultural production, for instance" (Krieger 2015). Writing for *Scientific American*, Douglas Fischer also raises the "central question: Who sets the thermostat?" (Fischer 2009). He cites to Ken Caldeira in describing the risks:

> "My biggest fear is that we're getting into the controls of the planet," said Calgary's Keith, "where one part of the world wants to run the planet different than another. [. . .] If one tweaks the knob a different way than another—or adds one knob atop another—it could be a real disaster." (Fischer 2009)

As these examples illustrate, while the thermostat metaphor implies a sense of easiness, it is often employed in an ironic manner that points to the political difficulties that are inextricable from the technical feasibility.

Both the studies by Nerlich and Jaspal (2012) and Luokkanen et al. (2014) included discussion of medical metaphors, in which planet Earth is discursively treated as a living organism needing healing or protection from bodily harm, or mechanical metaphors in which the Earth is treated as a machine that can be fixed. The present corpus of news articles is consistent in including similar medical and mechanical metaphors. These include medical metaphors comparing albedo modification to "chemotherapy for the planet" (Nocera 2015; see also Specter 2012). One notable factor is the path through which certain metaphors are introduced or perpetuated within the discourse.

In the case of some of the medical analogies, certain influential individuals are often cited. The chemotherapy metaphor has been attributed to various elite academics involved in geoengineering research, including Hugh Hunt of Cambridge University (see Specter 2012) as well as Gernot Wagner (see Nocera 2015) and David Keith (Nerlich and Jaspal 2012, 139; Howell 2010) of Harvard University. Ken Caldeira, climate scientist at the Carnegie Institution for Science at Stanford University, is quoted as referring to geoengineering as "kind of a symptomatic relief [. . .] I'm thinking like morphine for the cancer patient" (Carr 2015). These medical metaphors introduced by members of the geoengineering clique rhetorically treat geoengineering as non-curative symptomatic relief but at the same time imply through the comparison that it may nevertheless be a legitimate and necessary standard of care to pursue.

Journalists and editorialists also contribute to these metaphors. For example, an article in the *Economist* (2010a) includes an extended medical analogy:

"Cooling might take the edge off the peak of a planetary fever, or perhaps buy time as emissions cuts begin to have the desired effects. But hazing is a complementary medicine, not an alternative one." Michael Specter writes in the *New Yorker*: "Many people see geoengineering as a false solution to an existential crisis—akin to encouraging a heart-attack patient to avoid exercise and continue to gobble fatty food while simply doubling his dose of Lipitor" (Specter 2012). While the medical analogies attributed to geoengineering researchers listed above tend to use the metaphors to indicate legitimacy (e.g., likened to chemotherapy for a cancer patient), those by journalists include presentation ranging from positive or neutral ("complementary medicine") to a negative or critical tone (as exemplified in Specter's Lipitor analogy).

In addition to the medical metaphors, proponents use mechanical metaphors and analogies to other technologies toward legitimizing pursuit of the technical pursuit of geoengineering. These analogies are distinct from the types of metaphors previously discussed. As opposed to being morphine or chemotherapy in a metaphorical sense, geoengineering is directly compared to existing technologies. David Keith, the most oft-cited scientist in the field, has also imparted mechanical metaphors toward the normalization of geoengineering. Keith is quoted in one *Washington Post* article as saying: "A muffler is a technological fix for the fact that the internal combustion engine is very noisy, and people don't have a problem with mufflers" (Achenbach 2015).

Taking this to another level strategically, geoengineering advocates use analogies to other technologies that were themselves controversial at one point in order to diminish the sense of novelty and the grand extent of global risk imbued in the pursuit of geoengineering. This is seen in the *Foreign Affairs* article entitled "The Geoengineering Option" (2009) by David G. Victor, M. Granger Morgan, Jay Apt, John Steinbruner, and Katharine Ricke. They write:

Assessing and managing the risks of geoengineering may not require radically different approaches from those used for other seemingly risky endeavors, such as genetic engineering (research on which was paused in the 1970s as scientists worked out useful regulatory systems), the construction and use of high-energy particle accelerators (which a few physicists suggest could create black holes that might swallow the earth), and the development of nanotechnology (which some worry could unleash self-replicating nanomachines that could reduce the world to "gray goo"). The option of eliminating risk altogether does not exist. Countries have kept smallpox samples on hand, along with samples of many other diseases, such as the Ebola and Marburg viruses, despite the danger of their inadvertent release. All of these are potentially dangerous endeavors that governments, with scientific support, have been able to manage for the greater good. (Victor et al. 2009, 75–76)

Victor et al. choose comparisons to technologies that had initially provoked deep concerns but which were largely dismissed after further development. Analogizing to contested technologies, which have experienced a trajectory of normalization, while pointing to the most extreme characterizations of risk (straw man arguments regarding "black holes" and "gray goo") that have been presumably discredited, helps the authors make a case to dismiss and trivialize concerns regarding albedo modification. This example is, of course, on the advocacy side of the spectrum of public discourse on geoengineering.

News articles more broadly employ comparison to other contested technologies such as nuclear technology and genetic engineering. For example, nuclear technology can be used toward positive, neutral, or negative effects in analogizing, and hence framing, geoengineering. In arguing that the risks of geoengineering can be contained through scientific norms, Victor et al. point toward the nuclear precedent as a positive analogy: "Scientists could be influential in creating these norms, just as nuclear scientists framed the options on nuclear testing and influenced pivotal governments during the Cold War" (Victor et al. 2009, 74). Employing a neutral nuclear analogy, Daniel Cressey in *Nature* quotes,

> Shobita Parthasarathy, a public-policy researcher at the University of Michigan, Ann Arbor, [who] says that the field urgently needs to agree on detailed rules for IP. [. . .] One possible solution, she says, is to develop a unique system for handling geoengineering patents, akin to the way that atomic-energy patents are controlled in the United States. (Cressey 2012)

(While the analogy itself is neutrally employed, such comparison to existing technologies may contribute to a normalizing effect in the overall framing of geoengineering.) Employing the nuclear analogy as a negative framing, an article in *Slate* posed the argument and leading question:

> We need global norms that take into account the uncertainty and serious risks that solar radiation management could pose. [. . .] If early experiments epically fail, will they be counterproductive to the technology over the long term, like the nuclear meltdown in Three Mile Island? (Venkataraman 2016)

As these examples demonstrate, the nuclear analogy has variously been used across the entire spectrum of positive, neutral, and negative framings.

Similarly, analogy to genetic engineering is also employed toward various framings. As an example of a positive framing promoting consideration of geoengineering (along with genetic engineering), Andrew Revkin writes for the *New York Times* environmental opinion pages: "Walling off this arena makes as little sense as talking about feeding some nine billion people on a still-biodiverse planet without technology, including genetic engineering" (Revkin 2016). Through firm assertion of necessity, he dismisses arguments that would prima facie write off either of these controversial technologies.

As was seen with the nuclear analogy, analogizing geoengineering to genetic engineering is also at times used neutrally to discuss the formation of norms as a potential safeguard. For example, writing for *Slate*, Bina Venkataraman states:

> A recent global summit on gene editing technologies hosted by national scientific councils from the United States, the United Kingdom, and China could provide a model for how policymakers, ethicists, scientists, and the public can set boundaries on the use of technologies with unknown and intergenerational consequences. (Venkataraman 2016)

An article in the *Economist* (2010b) is premised on an extended analogy to genetic engineering to discuss the importance and possibility of regulation of the technology. Again, using these analogies as precedents for possible regulation, norms, and safeguards can help to normalize and diminish the sense of risks involved with novel emerging technologies.

Both metaphors and analogies are important tools for framing the presentation of the nascent concepts of geoengineering to the public. As seen in the examples presented here, individual metaphors or analogies can be used toward constructing positive, neutral, or negative framing. The wider corpus of articles examined here reaffirmed some of the findings of earlier studies, such as the application of medical and mechanical metaphors, while expanding upon these findings to identify the diversity of framings, including the use of positive, neutral, and negative framings that can be constructed even through the exact same analogies. Analysis of this broad and diverse corpus also introduced the importance of direct analogies in contributing to various framings and potential normalization of proposed geoengineering technologies

CONCLUSION

News media presentation of geoengineering includes a filtering through of scientific rhetoric and discursive framings that originated within the scientific community, such as from the statements of scientists or science policy reports. In addition to curating and repackaging scientific discursive framings and narratives for public consumption, journalistic media also insert their own framings and presentation styles that affect how the concepts are portrayed. As discussed in this chapter, the trends seen in journalistic articles on geoengineering include, for example, an emphasis on controversy as well as directing attention to "decoy" geoengineering proposals. While driven, in part, by substance (i.e., geoengineering *is* controversial), the ways in which these framings and presentation styles are used reflect broader trends in journalism, such as the tendency to present two sides of a debate, emphasizing points of contention, or the journalistic intrigue that comes from emphasizing the more sensational elements of a technological field. Through both the particular relaying of scientific discursive framings as well as through journalistic framings and styles inherent to the genre, the ways in which general audience media present geoengineering affect the public discourse on the topic, including how the concepts are introduced to new audiences.

As this and the previous chapter have shown, there exist multiple tensions within the public coverage of geoengineering. First, news media emphasize both the purported legitimation of the field while at the same time highlighting a sense of controversy within and surrounding it. In this coverage, certain voices are more influential than others, as seen in journalistic trends favoring citation of particular geoengineering experts and commentators. General audience discourse in news media also highlights the extreme inherent risks and, at times, absurdities of proposed technologies. Some of these extreme options act as decoy proposals within an implied range of options that facilitate homing in on more favored proposals. While, on the one hand, emphasizing absurdity and risk of geoengineering proposals, on the other hand, media often indicate that the failure of mitigation is paving the way for geoengineering. Moreover, while generally recognizing the importance of continued mitigation efforts, as articulated by climate experts including geoengineering advocates, popular media can falsely imply geoengineering exists as a mitigation alternative.

In addition to these original findings, this research expands upon existing understandings of how metaphors can be used toward various framings in geoengineering discourse. While reaffirming certain findings of other studies, the present discursive study both expands upon and deepens the scope of analysis through contextual consideration of such metaphors in a broad corpus and identifying how the same category of analogy or metaphor can be used toward divergent purposes. Furthermore, the element of analogies to existing technologies is appended to the conceptualization of metaphors in geoengineering discourse.

News media publications of geoengineering, of course, encompass a broad range of voices, vantage points, opinions, framings, and highlighted facts. Notwithstanding the challenges of assimilating the disparate themes inherent in such a wide range of reporting and editorializing on the topic, through in-depth discourse analysis of a broad range of articles within the universe of geoengineering-focused news articles, this study has aimed to contribute toward better illuminating the ways in which this potentially world-changing technology is presented to the public.

ADDENDUM: BEYOND 2016

This analysis has covered news media articles published in the years 1991–2016, which were the formative years in terms of geoengineering moving from obscurity to a topic of public discourse. During this time certain narratives and trends developed and conventionalized in the public discourse of geoengineering, as explored in these two chapters.

In the years that followed, there is some indication that at least a subset of news articles published around 2018 began casting geoengineering in a more benevolent light and emphasizing the role it may have to play in confronting climate change. For example, an article posted in the Climate News Network opens with the statement: "Progress to deploy solar engineering, experimental technology designed to protect the world against the impact of the changing climate, must pause, a former United Nations climate expert says, arguing that governments need to create 'effective guardrails' against any unforeseen risks" (Kirby 2018). While this excerpt references "unforeseen risks," what is unique about it is the otherwise bold and unreservedly positive language regarding solar geoengineering being "designed to protect the world" (Kirby 2018).

An article published in the Global Development section of the *Guardian* in spring of 2018 starts with the familiar science-fiction trope but then makes a significant pivot: "It sounds like the stuff of science fiction: the creation, using balloons or jets, of a manmade atmospheric sunshade to shield the most vulnerable countries in the global south against the worst effects of global warming" (Beaumont 2018). A benevolent imagery is evoked in the description of a "sunshade to shield the most vulnerable." This article is also notable in its title: "Scientists suggest a giant sunshade in the sky could solve global warming," which implies that albedo modification could stand alone as a solution rather than as a supplement or stopgap to emissions reductions (Beaumont 2018). In fact, this article, unlike comparable earlier publications, does not mention traditional mitigation at all.

Advancements in CDR also received news coverage in the intervening years. For example, the science-fiction trope was recycled to discuss emergent corporate support for CDR technologies. Brad Plumer and Christopher Flavelle's (2021) *New York Times* article, entitled "Businesses Aim to Pull Greenhouse Gases From the Air. It's a Gamble," included a caption reading: "A surge of corporate money could soon transform carbon removal from science fiction to reality. But there are risks: The very idea could offer industry an excuse to maintain dangerous habits." This article acknowledges the historic discourse around CDR technology and its chances for implementation as being primarily evaluated on a basis of cost, opening with the statement: "Using technology to suck carbon dioxide out of the sky has long been dismissed as an impractical way to fight climate change—physically possible, but far too expensive to be of much use." Pivoting from this, the article focuses on corporate businesses that have embraced CDR, as in "large companies facing pressure to act on climate" (Plumer and Flavelle 2021). Facing this public pressure, a number of industries with high carbon footprints, including oil companies, airlines, technology companies, and e-commerce companies, have begun investing in CDR technologies.

Climate news also continues to reflect the lived experience of climate change. For example, in 2021 news coverage included significant attention to climate disasters playing out in live-time and their connections to climate change. Climate change–related disasters prevalent in 2021, and reflected accordingly in news media, included unprecedented flooding (e.g., Victor 2021; Watts 2021), extreme wildfires (e.g., Canon 2021; Fountain 2021b; Jandt and York 2021; Wigglesworth 2021), and the increased frequency, range, and intensity of hurricanes (e.g., Samenow et al. 2021). Nestled appropriately among the news headlines of all the current climate disasters currently unfolding, coverage also included discussion of the 2021 IPCC Climate Report and its findings that "the world cannot avoid some devastating impacts of climate change, but that there is still a narrow window to keep the devastation from getting even worse" (Fountain 2021a).

These examples not only indicate, to a certain extent, a continued trajectory of climate coverage and geoengineering themes but also hint at some emerging trends in geoengineering coverage, especially in terms of benevolent descriptions and metaphors in regard to solar geoengineering. Time will tell whether these become more common. Overall, this is an area due for more research to affirm to what extent the trends and conventions continued from the previous 25 years and to what extent there was aberration from the trend lines, especially because the years 2017–2020 were unique in terms of climate discourse, especially in the United States with the tumultuous politics around climate change as the Trump administration made concerted efforts to erase climate change from the Federal vocabulary and roll back mitigation policies.

As a result, much coverage of climate change from 2017 to 2020 included political coverage of the United States' rollbacks to climate measures and commitments. Parallel to this, the Trump administration's positive views on geoengineering received some press attention. For example, Martin Lukacs writes for the *Guardian* in 2017 that "under the Trump administration, enthusiasm appears to be growing for the controversial technology of solar geo-engineering, which aims to spray sulphate particles into the atmosphere to reflect the sun's radiation back to space and decrease the temperature of Earth" (Lukacs 2017). Similarly, support for geoengineering had also been growing in the United States Congress, as will be analyzed in the following chapter.

NOTE

1. Recall that this larger universe of articles included all English-language articles within the LexisNexis database that had any relevant mention of geoengineering, while the core corpus is exclusive to articles focused primarily on geoengineering.

Section IV

CONGRESSIONAL HEARINGS

Chapter 6

Geoengineering in the Political Sphere

Congressional Hearings, 2009–2017

INTRODUCTION TO U.S. CONGRESSIONAL HEARINGS ON GEOENGINEERING

The United States House of Representatives includes 20 standing committees and numerous sub-committees therein. Each committee "considers bills and issues and recommends measures for consideration by the House" (United States House of Representatives 2017). Among their activities, committees "frequently hold hearings to receive testimony from individuals not on the committee" (Office of the Clerk 2018). According to the U.S. Government Publishing Office (1999), the purpose of congressional hearings is

> to obtain information and opinions on proposed legislation, conduct an investigation, or evaluate/oversee the activities of a government department or the implementation of a Federal law. In addition, hearings may also be purely exploratory in nature, providing testimony and data about topics of current interest.

The United States Congress held four hearings on the topic of geoengineering between November 2009 and November 2017. These hearings were before the U.S. House of Representatives Committee on Science and Technology, later reconstituted as the Committee on Science, Space, and Technology, as well as constituent subcommittees on Energy and Environment (see table 6.1). Each hearing considered a specific aspect of geoengineering. The first hearing in November 2009 was entitled: "Geoengineering: Assessing the Implications of Large-Scale Climate Intervention." The second hearing occurred in February 2010 and was entitled: "Geoengineering II: The Scientific Basis and Engineering Challenges." The third hearing occurred soon thereafter, in March 2010, and was entitled: "Geoengineering III: Domestic and International Research Governance." After a significant time gap, the fourth hearing occurred in November 2017, entitled "Geoengineering: Innovation, Research, and Technology." See table 6.1 for an overview of some relevant details relating to the four geoengineering hearings. These four hearings are the focus of analysis in this chapter.[1]

The panels for the four hearings were comprised of four or five testifying witnesses with experience in various aspects of geoengineering research. Each of these panels included two to four witnesses from academic institutions in addition to one or two affiliated with either a national laboratory or a think tank. The March 2010 hearing supplemented this witness pattern with one representative from the Government Accountability Office (GAO) and the unique inclusion of a member of Parliament from the House of Commons in the United Kingdom, as there was a joint inquiry on geoengineering between the counterpart science committees in the U.S. House of Representatives and the U.K. House of Commons. Table 6.2 provides a detailed list of panelists. In terms of committee member participation, as seen in table 6.1, the first and the fourth hearings each included statements or questions from 11 congressional representatives, while the second and third hearings were sparsely attended and included statements and questions from three and four congressional representatives respectively.

External factors beyond the scope of the Committee influence the attendance, course, and progression of these events. As Hugh Mehan pointed out in his study of "social structure and power as an interactional process" within schools, "circumstances which originate outside the institution ['distal circumstances'] interact with circumstances which originate within it ['proximal circumstances'] to influence the course of interaction and the work of the formal organization" (Mehan 1987, 291, 293). Such distal and proximal circumstances affected the proceedings and interactions studied within these congressional hearings. As will be discussed further, political disagreements external or tangential to the specific subject matter of the hearings influenced the interactions within these geoengineering hearings. Furthermore, completely separate policy issues that overlapped temporally with these congressional hearings affected the proceedings, competing for the time and attention of congressional representatives. Particularly, votes and political debates taking place within the House on other topical matters affected attendance and performance at the geoengineering hearings.[2]

When votes were scheduled to occur in the House of Representatives, the hearings were, of necessity, more rushed so that the members could leave to cast their votes. The Committee Chair's time management of the hearing proceedings was clearly affected by the voting schedule. For example, partway through the first hearing (Congressional Hearing 2009a), Chairman Bart

Table 6.1 Overview of U.S. House of Representatives Hearings on the Subject of Geoengineering

Hearing Title	Date	Convening Body	Presiding Chair	# Witnesses	# Members
Geoengineering: Assessing the Implications of Large-Scale Climate Intervention	November 5, 2009	Committee on Science and Technology	Bart Gordon (D-TN)	5	11
Geoengineering II: The Scientific Basis and Engineering Challenges	February 4, 2010	Subcommittee on Energy and Environment (of the Committee on Science and Technology)	Brian Baird (D-WA)	4	3
Geoengineering III: Domestic and International Research Governance	March 18, 2010	Committee on Science and Technology	Bart Gordon (D-TN)	5	4
Geoengineering: Innovation, Research, and Technology	November 8, 2017	Subcommittee on Environment and Subcommittee on Energy (both of the Committee on Science, Space, and Technology)	Andy Biggs (R-AZ)	4	11

Gordon declared: "I am going to be a little more strict because we are going to votes, unfortunately, in a few minutes" to which Vernon Ehlers, Republican representative from Michigan, replied: "It is so amazing how the clock runs so much faster when it is my time." At the second hearing (Congressional Hearing 2010a), because of timing of votes conflicting with the timing of the hearing, the witnesses were introduced, and some statements were made before the hearing was officially called to order, with opening verbal statements foregone (written statements were submitted into the record). An hour recess was held in the middle for votes.

The third hearing (Congressional Hearing 2010b) was also affected by votes, with the witnesses being warned multiple times that imminent voting may necessitate a recess or conclusion of the hearing. During witness questioning, Chairman Gordon's questions were affected by this external time constraint. Chairman Gordon asked the panel their thoughts on "what agency or agencies would be the appropriate vehicles for this type of research" (Congressional Hearing 2010b). However, after one panel member responded to this question, Chairman Gordon interjected to say: "We are being called for votes, so let me just ask, I would assume everyone concurs with that, unless you have a suggestion of something specific. Otherwise, is there anyone that has anything else?" (Congressional Hearing 2010b). As it turned out, the other three members of the panel all wanted to respond to the original question regarding agencies and were allowed to make brief statements, after which Chairman Gordon ended the hearing, stating: "As I said, we, our votes are on their way right now, so let me thank all of our witnesses for being here" (Congressional Hearing 2010b). As demonstrated here, external factors and constraints affected the course of the hearing and in this example limited a discussion substantially relevant to government oversight of geoengineering research, which would have likely continued in more depth had there been additional time.

This set of four hearings provides an opportunity to examine discourse in interaction, including the presentation and reception of geoengineering narratives and framing between two categories of elite actors, scientists, and congressional representatives. The witnesses, mostly scientists with significant experience researching geoengineering concepts, provide testimony. The congressional committee members receive the same testimony differently with clear divisions in how they interpret it, especially along party lines, but with some variance and change over time. Consistent with the trends found within other spheres of discourse, the progression of these four congressional hearings, occurring over an eight-year period, illustrates the increasing mainstreaming over time of geoengineering as a serious consideration. Of course, this setting of policy makers considering geoengineering not only reflects views on the topic but is intended, as per the purpose of congressional hearings, to affect the trajectory of policy on it.

PREMISES ARTICULATED FOR GEOENGINEERING
AND THE CONGRESSIONAL HEARINGS

As discussed in the previous chapters, as well as in other studies (e.g., Bellamy 2013; Corner, Parkhill and Pidgeon 2011; Nerlich and Jaspal 2012), two central framings within geoengineering discourse are the emergency/ catastrophe framing and the insufficiency of mitigation framing. These two framings were clearly identifiable within the charter for the first geoengineering hearing, which articulated the premise for both the hearing and for geoengineering as a possible response to climate change (Congressional Hearing 2009a). Exemplifying the insufficiency framing, the charter stated: "Many in the international climate community hold that even the most aggressive achievable emissions reductions targets will not result in the avoidance of adverse impacts of climate change and ocean acidification" (Congressional Hearing 2009a, 4). It continued on with an instantiation of the climate emergency framing to state: "Further complicating these projections is the possibility of non-linear, 'runaway' environmental reactions to climate change. Two such reactions that would amount to climate emergencies are rapidly melting sea ice and sudden thawing of Arctic permafrost" (Congressional Hearing 2009a). Thus, per the charter: "It is for these reasons that geoengineering activities are considered by some climate experts and policymakers to be [a] potential 'emergency tool' in a much broader long-term and slower acting global program of climate change mitigation and adaptation strategies" (Congressional Hearing 2009a). In this way, the premise of geoengineering and the justification for holding hearings on it were defined by the common narratives of insufficient mitigation and climate "emergency."

The initial hearing charter also engaged with the notion that geoengineering has recently moved toward mainstream consideration. The charter stated: "Scientific hypotheses resembling geoengineering were published as early as the mid 20th century, but serious consideration of the topic has only begun in the last few years" (Congressional Hearing 2009a, 4). This assertion of geoengineering now being subject to "serious consideration" presents it as an appropriate topic for the committee's consideration, which in itself signifies a step in the direction of mainstreaming. Moreover, a list of respected mainstream organizations that have given attention to the idea of geoengineering is used as evidence of its move from obscurity to mainstream consideration. This list, underscoring legitimate ("serious") consideration of geoengineering, includes the National Academy of Sciences, the Intergovernmental Panel on Climate Change, the U.S. Department of Energy, NASA, DARPA, the NSF, and the Royal Society (Congressional Hearing 2009a, 4–5). In this way, there is a circular process snowballing the assertion that geoengineering is now mainstream based upon the elite institutions examining it, as each new institution points to those before it to justify the seriousness of the topic. Following these hearings, the House Committee on Science and Technology itself could be added to this list of elite institutions indicating legitimacy of geoengineering through their engagement with the topic.

In addition to introducing geoengineering concepts and explaining the premises for their consideration, the background section of the hearing charter also explored various risks and challenges that would be involved with potential geoengineering. It drew significantly on the Royal Society report, which was the definitive geoengineering report to date in 2009. Like the science policy reports, this initial charter advocated research but raised caveats regarding deployment. It drew comparisons to nuclear weapons testing and the history of weather modification attempts as "Analogous Government Initiatives" characterized by significant risks and uncertainties, stating these technologies "display a number of similarities to geoengineering, including the difficulties of levying cost-benefit analyses of their impacts, uncertain ecological impacts, an unknown geographic scope of impact, and potential intra- and intergovernmental liability issues" (Congressional Hearing 2009a, 9). The nuclear testing and weather modification analogies were used as examples of precedential government initiatives that constitute domestic technological programs with international repercussions "incurred without international consent" (Congressional Hearing 2009a, 9).

In Chairman Bart Gordon's opening statement at this first hearing, he specified that his decision to hold the hearing did not indicate "an endorsement of any geoengineering activity" but said that the topic "requires very careful consideration" as a potential "stopgap" measure or response to "a climate emergency" (Gordon, Congressional Hearing 2009a, 11–12). Chairman Gordon stated: "We must get ahead of geoengineering before it gets ahead of us, or worse, before we find ourselves in a climate emergency with inadequate information as to the full range of options" (Gordon, Congressional Hearing 2009a, 12). As exemplified in Chairman Gordon's statement, at these hearings, consistent with other genres of geoengineering discourse discussed in the previous chapters, the catastrophe/emergency framing and the stopgap/buy-time framing are two primary narratives used in support of geoengineering research and consideration. However, like science policy reports, he also emphasized that "nothing should stop us from pursuing aggressive long-term domestic and global strategies for achieving deep reductions in greenhouse gas emissions" (Congressional Hearing 2009a, 11–12). For the most part, this set the tone for the Democrat Committee members' discussion of geoengineering, which evolved somewhat over time but without ever losing this core duality of open but cautious consideration of geoengineering paired with a reiteration of the primacy of emissions mitigation. As will be discussed, it is also closely aligned with the core messages of the witnesses who provide testimony at the four hearings. In contrast, the Republican position on geoengineering makes a dramatic transition over time.

Premises and Partisan Performances

Due to an interplay of structural elements, external factors, and specific performances of politicians and witnesses, the four congressional hearings on geoengineering varied in content, tone, and participation but always within the confines of certain institutionalized expectations. Despite the broad range of participation by Committee members, varying from three to eleven participating members, the formal elements of congressional hearings were steadfastly observed, with members invariably stating "Thank you, Mr. Chairman" as they began their timed five-minute periods for opening statements, asking questions of the witnesses, or formally requesting certain documents be included in the official record.

Other elements of the hearings, including partisan performances, were less formal, although no less entrenched. Within the structure of the hearings, politicians tried to claim the framing and representation of the issues being discussed. There were certain narratives and performances members of each party felt compelled to play out. The Democratic members of the Committee consistently and repeatedly reiterated at each hearing the existence of anthropogenic climate change and the primacy of mitigation and adaptation in addressing it. For the first three hearings, prominent Republicans displayed an opposing performance, questioning the importance or relevance of anthropogenic climate change and thus the premise for the hearings at all. At the fourth hearing, an interesting shift occurred in which Republicans enthusiastically embraced geoengineering. This dramatic shift was facilitated through adopting a strategy of rhetorically decoupling geoengineering from climate change as to minimize the cognitive dissonance of this newfound enthusiasm for solar geoengineering in spite of the party position on climate change.

In articulating the purpose of the hearing, Committee Chairman Andy Biggs' opening statement to the fourth hearing in November 2017 explicitly specified: "The purpose of this hearing is to discuss the viability of geoengineering . . . the hearing is not a platform to further the debate about climate change; we've had lots of that this session" (Congressional Hearing 2017b). Since geoengineering is intricately related to climate change, it would not be expected that the two could be separated. Yet, during this hearing, Chairman Biggs and his fellow Republican members noticeably avoided the topic of "climate change," with most Republicans avoiding the concept and term entirely or obfuscating the concept when it could not be avoided. For example, Randy Weber, a Texas Republican and the chair of the Energy Subcommittee, enthused over the "bright" prospect of geoengineering while obfuscating its relationship to climate change. In his opening remarks to this hearing, he paused for a moment before clearly enunciating a demarcation between the topics of geoengineering and climate change:

> *If we put aside the debates about climate change,* we can support innovations in science that create a better prospect for future generations. The federal government should prioritize *this* kind of basic research, so we can not only understand the science of geoengineering, but hopefully partner with the private sector to develop technology to mitigate changes in climate. When the government supports basic research, everyone has the opportunity to access the fundamental knowledge that can lead to the development of future technologies. The future is bright for geoengineering. (Congressional Hearing 2017, italics based on verbal emphasis)

Within this statement, two conflicting elements coexist: advocacy for pursuit of geoengineering alongside a dismissal of the relevance of climate change upon which geoengineering is premised. Weber suggests that geoengineering may be an area of political agreement under the condition that it is separated from climate change. By stating, "If we put aside the debates about climate change," Weber proposed decoupling of geoengineering from climate change while at the same time perpetuating the partisan argument that there are "debates" regarding the existence of climate change (Congressional Hearing 2017b).

In contrast, the ranking Democratic member of the Committee at this November 2017 hearing, Suzanne Bonamici of Oregon, couched her opening statements in terms of the importance of addressing climate change, emphasizing that mitigation and adaptation must be the first avenues irrespective of geoengineering: "Even with geoengineering, our first and primary actions to address climate change must be mitigation and adaptation strategies" (Congressional Hearing 2017b). Bonamici, like other Democrats on the Committee, reiterated the existence of anthropogenic climate change:

> Our climate is changing and the warming trends observed over the last 100 years are primarily caused by human activities, specifically the emission of greenhouse gases. In fact, this is one of the most prominent findings in the Climate Science Special Report [Fourth National Climate Assessment, a government report released November 2017, shortly before this hearing occurred]. This report unequivocally lays out the need to reduce carbon dioxide emissions to prevent long term warming and short term climate change. (Congressional Hearing 2017b)

This reiteration of basic facts of anthropogenic climate change is recurrent among Democrats on the committee.

Also at the fourth hearing, Marc Veasey, a Democrat from Texas, made the seemingly unequivocal statement in his opening remarks that

> despite the numerous claims, geoengineering is not the answer to 150 years of polluting our planet at an unsustainable rate [. . .] we have to get our priorities straight and mitigation and adaption must be part of the top priorities. (Congressional Hearing 2017b)

Yet, despite this position, he came around to the pursuit of geoengineering, stating: "The long-term nature of this challenge [climate change] is the reason we need to investigate every possible solution in addition to implementing mitigation and adaptation strategies" (Congressional Hearing 2017b).

Jerry McNerney, Democrat from California, has been one of the strongest supporters of congressional consideration of geoengineering. McNerney advocated for holding the November 2017 hearing and subsequently, in December 2017, introduced a bill in support of geoengineering research to Congress. His bill, H.R. 4586, the Geoengineering Research Evaluation Act, "would provide for a federal commitment to the creation of a geoengineering research agenda and an assessment of the potential risks of geoengineering practices" (McNerney 2017a). It would require the Federal government to "contract with the National Academies to conduct a study and develop a report recommending a research agenda for advancing understanding of albedo modification strategies that involve atmospheric interventions" (McNerney 2017b).[3] In 2019, Jerry McNerney introduced another geoengineering-related bill, H.R. 5519, the Atmospheric Climate Intervention Research Act, which would expand "the directive of the Office of Oceanic and Atmospheric Research within the National Oceanic and Atmospheric Administration to include climate intervention research, including research regarding the effects of proposed interventions in the stratosphere and in cloud aerosol processes" (McNerney 2019).[4]

Within the hearing, like his Democratic colleagues, McNerney reiterated facts of anthropogenic climate change: "Climate change is happening and the effects are accelerating faster than the scientific models predict [. . .] meanwhile carbon concentration in the atmosphere is continuing to increase" (Congressional Hearing 2017b). Also like his colleagues, McNerney reiterated the importance of mitigation and adaptation: "no matter what, it is absolutely critical to reduce carbon emissions and prepare for the changes coming" (Congressional Hearing 2017b).

In addition to reiterating the importance of mitigation and adaptation, McNerney also employed a very clear use of the catastrophe framing in support of research toward geoengineering (Congressional Hearing 2017b). In articulating his reasoning regarding geoengineering, McNerney implicitly referenced the challenge of latency in regard to GHGs, which includes the lag-time between GHG emissions and the resulting climatic effects, plus the fact that GHGs persist in the atmosphere for long time periods. He stated:

> We are committed to significant change. The unknown is how much change we are committed to and how fast it will take place. It is not known if we are committed to truly catastrophic change with the current policies or not.

He then laid out geoengineering as a tool for addressing potentially catastrophic climatic changes: "The changes we are committed to may be so strong that we need to know what can be done to prevent utter catastrophe" (Congressional Hearing 2017b).

Similarly, at an earlier hearing, Brian Baird, Democrat from Washington, combined the narratives of the primacy of emissions abatement with catastrophe and Plan B framing. Baird stated "Without question, our first priority is to reduce the production of global greenhouse gas emissions. However, as I said, if such reductions achieve too little, too late, there may be a need to consider a plan B" (Congressional Hearing 2010a).

Democrats within the Committee, reflecting the role of their party more broadly in government and consistent with the Democratic Party's 2016 platform, took on the role of vigilant reiteration of the reality and significance of climate change and promoting mitigation and adaptation policy. At the November 2017 hearing, each Democrat who spoke once again reiterated the importance of prioritizing mitigation and adaptation, while (often reluctantly) accepting the premise of geoengineering research. Despite a reluctance among Democrats to show enthusiasm for geoengineering, the internal logic of their position on climate change and the evident frustration expressed by some members regarding the failure to enact meaningful mitigation policy, allowed an opening for the notion. This stance was succinctly put by Marc Veasey, as mentioned above, in saying "We need to investigate every possible solution in addition to implementing mitigation and adaptation strategies" (Congressional Hearing 2017b).

On the other side of the aisle, Republicans made a remarkably drastic turn-around on their treatment of geoengineering. After three hearings of dismissing the premise and relevance of the hearings, in the fourth hearing, Republicans (especially those members outspoken on denying climate change) came to embrace geoengineering with a newfound enthusiasm. While surprising, this pattern is not unique. There has been a trend of erstwhile climate change deniers becoming geoengineering advocates in other instances as well (Hamilton 2013a, 76–77, 85, 98–99, 129). For politicians, however, to reconcile the paradox of supporting geoengineering while simultaneously disputing the importance of climate change, requires a redefining of geoengineering.

The official records for the first and third hearings characterized geoengineering as "the deliberate large-scale modification of the earth's climate systems for the purposes of counteracting climate change" (Congressional Hearing 2009a; Congressional Hearing 2010b). At the fourth hearing, Suzanne Bonamici, the ranking Democrat on the Committee, defined geoengineering as follows: "Geoengineering is a set of climate interventions that aim to manipulate our climate, to either remove greenhouse gases from the atmosphere or reduce the amount of sunlight absorbed by the Earth" (Congressional Hearing 2017b). She then proceeded to contextualize it in relation to other climate change policy options:

> Now some may argue that geoengineering is a way to use technology to bypass important mitigation and adaptation strategies that address the impacts of climate change, but even with geoengineering, our first and primary actions to address climate change must be mitigation and adaptation strategies. (Bonamici, Congressional Hearing 2017b)

In contrast, Chairman Andy Biggs, a Republican from Arizona, stated: "In its simplest terms, geoengineering is the concept of using scientific understanding to alter the atmosphere in a way that produces positive outcomes and results" (Congressional Hearing 2017b). In this way, he defined geoengineering without reference to climate change. This redefining of geoengineering allowed for Biggs and other Republican members of the committee to support geoengineering without acknowledging or accepting the reality of climate change.

This constituted a remarkable shift from the earlier hearings on geoengineering to the most recent one. The first three hearings each included at least one Republican Committee member rejecting geoengineering in absolute terms, in relationship to the rejection of climate change. The fourth hearing in November 2017, however, despite having a high degree of participation with as many Republicans speaking as at any of the hearings, included no statements of absolute rejection of the pursuit of geoengineering. This change is facilitated by recasting geoengineering to be defined without reference to climate change. It also portrays a manifestation of the trajectory of geoengineering toward mainstream consideration as has occurred in the eight-year period between the first and fourth hearings.

FORMS OF SKEPTICISM ENACTED AT GEOENGINEERING HEARINGS

In the first three hearings, individual Republican members employed one of two distinct styles in expressing their rejection of climate-change science and geoengineering technologies. These two contrasting rhetorical styles can be characterized as adversarial or collegial.

Dana Rohrabacher, Republican from California, clearly represented the adversarial style, characterized by confrontational statements and hostile questioning of the witnesses. At each of the first two hearings, Rohrabacher enacted a sort of script that he and his colleagues even indicated had come to be expected of him at climate-related hearings. The repetition of a script is suggested by the way in which Rohrabacher opened his allotted speaking time at the first of the geoengineering hearings:

> Thank you very much, Mr. Chairman, and no hearing like this would be fulfilled without my adding a list at this point of 100 top scientists from around the world who are very skeptical of the very fact that global warming exists at all, but I would like to submit that for the record at this time. (Congressional Hearing 2009a)

By stating that "no hearing would be fulfilled without" his introducing skepticism regarding the existence of climate change, Rohrabacher suggests that he regarded this as his primary role within such proceedings.

After submitting his list, he went on to question geoengineering based upon his questioning of climate science:

> There you go. Let me just note that there is ample reason for us to question whether or not things that are being suggested today are really needed because there is reason to question whether there is global warming, considering the fact that it has gotten—it is not gotten warmer for the last nine years, and the Arctic polar cap is now refreezing for the last two years. But that argument isn't what today's hearing is about, so I will just make sure that that is on the record and in people's minds when looking at some of these suggestions. (Congressional Hearing 2009a)

Rohrabacher's treatment of climate change follows the pattern, identified by Oreskes and Conway (2010), of sowing doubt regarding climate change through indicating a continuance of debate over its existence and questioning climate science, especially through emphasizing and mischaracterizing the existence of (normal) scientific uncertainty. At each of the hearings attended by Rohrabacher, this form of contesting the veracity of climate change and submitting the list of climate skeptics, was followed by adversarial interactions with the witnesses, in which he focused on problematizing the existence of climate change and mitigation concepts.

During the first hearing, the witnesses seemed taken aback by the line of questioning and attempted neutral responses steering the discussion back to the topic of geoengineering. Early in his questioning during this first hearing, Rohrabacher reacted negatively when Alan Robock corrected an inaccurate statement that Rohrabacher had made. This led to hostile questioning from the Representative and a response from Robock in which he attempted to maintain a calm tone of scientific authority, answering the questions based on scientific knowledge and practical considerations despite the shift from information-gathering toward antagonistic questioning:

Mr. ROHRABACHER: Let me ask about some of the specific suggestions. I . . . understand at 9/11 when they grounded all the airplanes that it actually increased the temperature of the planet, is that right? And thus——

Dr. ROBOCK: Excuse me, that is not correct.

Mr. ROHRABACHER: It is not correct?

Dr. ROBOCK: There was one study that showed that without clouds from contrails that the diurnal cycle of temperature went up, that the daily temperature went up, the nighttime temperature went down, but that was later disproven. It was shown that was just part of natural weather variabilities. So that wasn't a very—

Mr. ROHRABACHER: Let me note that every time it doesn't fit into the global warming theory, it becomes natural variability but when it does fit in, it becomes proof that there is global warming. Let me ask you this. That really wasn't then? Does anyone else have another opinion of vapor trails, by the way? So we have learned today that we really just have—and am I misreading you by suggesting that you, too, are part of the group that believes in global warming that would like to restrict air travel or try to find ways of eliminating frequent flyer miles? We know you don't want us to eat steak now. Are we also not going to be able to fly on airplanes?

Dr. ROBOCK: Airplanes are one of the sources of emissions. If they use biodiesel and it recycles the fuel, then it wouldn't be part of the problem. But indeed, if we—we can do some emissions of CO_2. We don't have to—these mobile transportation sources are very hard to retrofit on airplanes. With cars, you can, of course, generate electricity with wind and solar, but airplanes, we still have to keep flying and we can live with a little bit of CO_2 emission if we deal with other sources. (Congressional Hearing 2009a)

This interaction began with Rohrabacher starting down a line of argumentation, which seemed to be aimed at challenging the concept of albedo modification. This challenge was based upon the notion that planes being grounded in response to the terrorist attacks of September 11, 2001 led to increased temperature rather than decreasing it. If it were true that temperatures increased in response to the anomalous reduction in contrail emissions in the stratosphere, it would seem to challenge the premise of proposed forms of albedo modification, particularly stratospheric aerosols. However, before Representative Rohrabacher was able to articulate his argument, Dr. Robock interjected to clarify the scientific findings that he presumed were being used to inform Rohrabacher's claims and to indicate that these findings, and hence the premise of Rohrabacher's line of argumentation, had since been debunked, based on further scientific research. In this way, Robock challenged Rohrabacher's claims through problematizing his grounds for warranting those claims (Toulmin 1958), causing a shift in the tone and direction of the dialogue.

Being waylaid from his line of argumentation, Rohrabacher turned back to his original tactic of disputing climate change, despite having just indicated that he would move on from this topic because "that argument isn't what today's hearing is about" (Congressional Hearing 2009a). He returned to it, however, in the form of ad hominem attacks upon the panelist, with an accusatory tone when stating, "you, too, are part of the group that believes in global warming that would like to restrict air travel." While Alan Robock certainly believes in climate change and in the need to reduce carbon emissions, the tone and context of the communication was one of accusation. Following this, the representative ended with a seemingly rhetorical question ("Are we also not going to be able to fly on airplanes?"), which Robock forbearingly answered as if the question were genuinely of the information-gathering variety, referring to which modes of transportation are practical for decarbonization and conceding that air travel may be a particular exception to emissions abatement (Congressional Hearing 2009a).

In this interaction, after being thwarted from one line of argumentation, rather than pursuing rebuttal, Rohrabacher turned to questioning the legitimacy of climate scientists as a premise for questioning their authority. The representative suggested that Robock and other climate scientists are arbitrary in use of evidence and in distinguishing between natural weather variability and climate change. Following this statement, Rohrabacher proceeded to suggest that Robock is part of a group campaigning against air travel, although he did not articulate any particular grounds to warrant the claim. In these ways, Rohrabacher's rhetorical strategy seemed aimed at discrediting and denying Robock's neutrality and objectivity by implying some sort of ulterior motive underlying the climate scientist's statements on climate trends and the importance of mitigation. This constituted a direct challenge not only to Robock's statements but to his integrity as a witness and as a professional scientist. Robock answered in a measured way, stating facts and not directly responding to the attempted attack on his professional authority and integrity. Rather, he maintained a professional tone, giving scientifically-grounded, fact-based responses to the ostensible questions asked. Through projecting the demeanor of an impartial and unflappable scientist, Robock implicitly reasserted his credibility of scientific authority in response to its being rhetorically challenged by Representative Rohrabacher.

At the same hearing, following this exchange with Robock, there was a similar interaction between Representative Rohrabacher and Ken Caldeira. As discussed in previous chapters, climate scientist Ken Caldeira is a prominent and oft-cited geoengineering researcher. Rohrabacher's interaction with Caldeira again exemplifies the congressional member's approach to disputing the relevance of anthropogenic climate change from multiple angles and returning to the issue of his disapproval of actual or perceived suggestions that lifestyle changes may be necessary:

Mr. ROHRABACHER: [. . .] there are those who have realized—in the past there have been many times when that CO_2 content was enormously greater, wasn't that right? And during that time period there were lots of animals, like dinosaurs and lots of things growing, and the world seemed to be doing pretty good.

Dr. CALDEIRA: CO_2 concentrations were high in the past, and the biosphere flourished. And even if we disagree about what the threats are from climate change, and I think we do, that, you know, I don't think my house is going to burn down, but I buy fire insurance. And——

Mr. ROHRABACHER: But you don't tell your neighbor that he can't have steak or visit his kids in an airliner, and that is the point.

Dr. CALDEIRA: I don't——

Internally inconsistent with another claim, which he made at both this first hearing as well as the second hearing that the "tiny, miniscule amount" of CO_2 in the atmosphere may not have "anything to do with the changes in the climate" (Congressional Hearing 2010a), here Representative Rohrabacher referenced the age of the dinosaurs presumably to argue that higher CO_2 may not be a bad thing, but rather beneficial. When Caldeira tried to redirect the dialogue back to geoengineering with the house insurance analogy, Rohrabacher pivots back to his contention regarding climate-change mitigation, couched in second-person language ("you don't tell your neighbor that he can't have steak or visit his kids in an airliner"). The dialogue, diverted toward this question of lifestyle changes, continued like this:

Mr. ROHRABACHER: There are going to be changes. People have to understand, there are going to be huge changes in our lifestyle——

Dr. CALDEIRA: I don't——

Mr. ROHRABACHER: —if this nonsense is accepted.

Dr. CALDEIRA: I don't believe we are going to solve this problem by asking people to behave differently.

Mr. ROHRABACHER: Okay.

Dr. CALDEIRA: I think we are going to solve it by improving the systems that surround us. But to get back to my point, even if we don't believe that climate change will damage us, we have to say there is some risk. So then we have to say, well, how much should we invest to try to mitigate that risk.

Mr. ROHRABACHER: We are broke right now, and the bottom line is that we have very little to invest in theories that may or may not be correct, and we also have a lot of indication, just the fact that you are using the word climate change is a difference than what was used 10 years ago which was global warming. And most of us realize that is because people now are trying to hedge their bets so they can have these controls, whatever way the temperature goes.

Dr. CALDEIRA: No, I don't think that is true. You know——

Chairman GORDON: Time.

Mr. ROHRABACHER: Thank you very much.

Chairman GORDON: Speaking of dinosaurs, the time for Mr. Rohrabacher has run out, and we will need to proceed to——

Mr. ROHRABACHER: Thank you, Mr. Chairman. (Congressional Hearing 2009a)

In this interaction, Representative Rohrabacher repetitively interrupted the expert witness and used antagonistic language to dispute climate change, referring to it as "nonsense" and "theories that may or may not be correct" (Congressional Hearing 2009a). Moreover, this language is emblematic of the climate denial strategies, including sowing doubt regarding the veracity of climate change itself (see Oreskes and Conway 2010).

The designation of climate change as "theory" (let alone "nonsense") rather than "fact" is an important rhetorical move in making political space for alternative theories. It mirrors the rhetoric of anti-evolution creationism or "creation science" which insists that evolution is "only a theory" and therefore creationism should also be taught in science classes. In response to this line of argumentation, Stephen Jay Gould eloquently articulated:

> evolution *is* a theory (emphasis in original). It is also a fact. And facts and theories are different things, not rungs in a hierarchy of increasing certainty. Facts are the world's data. Theories are structures of ideas that explain and interpret facts. Facts do not go away when scientists debate rival theories to explain them. (Gould 1981)

Likewise, understanding and attempting to forecast the complex, multivariate interactive processes involved with climate change involves theories employing factual data. Climate scientists create models for various scenarios that may play out more or less accurately, but, in either case, the imperfections and uncertainties involved in modeling do not undermine the validity of the facts on which they are based. Moreover, theories remain, in the scientific meaning, logic and evidence-based systematic explanations for making sense of empirical reality even if theories are fine-tuned as new data emerge. However, the denialist argument instantiated by Rohrabacher is premised upon discrediting climate change on the basis of these misconceptions that theories are distinguished from fact and that modifications or uncertainty indicate problems rather than process.[5]

Furthermore, Rohrabacher's statements implied that proponents of climate mitigation policies employ deception (i.e., by changing terminology) to "hedge their bets" toward an end goal of emissions controls, although he did not articulate what interests he believed would underlie pursuit of emissions controls apart from purposes of climate mitigation (Congressional Hearing 2009a). The emission controls that he spoke of directly are those that fall into the category of lifestyle changes. In his interactions with both Robock and Caldeira, Rohrabacher rhetorically accused them of having an agenda to require some form of lifestyle changes, with the implication that this would be a suspect position to hold. In each of these two interactions, the respective scientists were speaking generally of climate science or the potential role of geoengineering, when the congressional representative insinuated that they had a personal agenda to change people's lifestyles. While Robock conceded that lifestyle components were one element of GHG mitigation (in discussion of reducing beef consumption), Caldeira indicated that he considered systems as more important than lifestyles for mitigation. In either case, these interactions demonstrate Representative Rohrabacher's multi-faceted approach toward challenging the existence of climate change through employing various rhetorical and argumentative strategies, including calling into question climate science itself as well as the authority and objectivity of the climate scientist witnesses by implying that they have a hidden agenda.

In both of these interactions, congressional member Rohrabacher made clear that one of his primary contentions with climate change was that addressing it would affect "our lifestyle," such as meat consumption and air travel (Congressional Hearing 2009a). Whereas during the later 2017 hearing prominent Republicans known for disputing climate change embraced albedo modification geoengineering as an option for bypassing lifestyle changes relevant to emissions abatement, Rohrabacher's 2009 position was demonstrative of absolute climate-change denial with internal consistency in its translation to dismissing geoengineering prima facie.

During the second hearing, the witnesses seemed more prepared for Rohrabacher's style of questioning and pushed back. Expert witnesses firmly disputed each of the representative's arguments, until Rohrabacher himself ended the discussion by stating "I see my time is up" (Congressional Hearing 2010a). Emphasizing the repetitive nature of Rohrabacher's performance, the response of the witnesses was commented upon by both the Democratic Subcommittee Chair and the Republican ranking member. Chairman Baird, Democrat from

Washington, commented: "I thank the gentlemen for their responses and want to commend you. Some of the arguments that Mr. Rohrabacher has made have been offered previously to panels of climate scientists without response, and I commend you for the response" (Congressional Hearing 2010a). Republican ranking member from South Carolina, Bob Inglis, also commented: "I am with Chairman Baird, I thank you for answering the question because quite often those questions do go—or those assertions go unchallenged and so very cogent explanation there" (Congressional Hearing 2010a).

These comments speak to Rohrabacher's repetitive, almost scripted, performance being a common and expected pattern. The representative repeated elements of his performance of climate-change denial in both of the first two geoengineering hearings but with very different results due to the distinct interactions with the witness panel. While Rohrabacher did not attend the subsequent two geoengineering hearings, at other hearings relevant to climate, including one later in the same congressional session entitled "Monitoring, Measurement, and Verification of Greenhouse Gas Emissions," Rohrabacher also enacted the rhetoric of climate skepticism and submitted to the official record a (different) list of scientists "who are in disagreement with the theory that greenhouse gases are" causing global warming (Congressional Hearing 2009b). In a hearing on "Science of Capture and Storage" in 2014, Rohrabacher again indicated that the premise of climate change was subject of ongoing "debate" and explicitly stated his position that "the concept of global warming is fraudulent and it has not been proven" (Congressional Hearing 2014). Representative Rohrabacher's consistent performances are, of course, emblematic of a dominant narrative perpetuated in politics and public discourse by those with interests in maintaining the status quo.

The geoengineering hearings were called for the purpose of information-gathering and the panelists were requested to testify as expert witnesses. Of the 24 unique representatives that spoke during the four hearings, only two took an antagonistic approach with the witnesses: Dana Rohrabacher, at the first and second hearings as discussed, and Adrian Smith at the first hearing. Adrian Smith, Republican of Nebraska spent his speaking time at the first geoengineering hearing hostilely asking the witnesses, who were all invited panelists due to their expertise on geoengineering, about whether they think beef, a major industry in Smith's district for which he is a strong proponent, is bad for the environment. Rohrabacher and Smith were anomalous in their interactions with the panelists in which they rhetorically challenged the credibility of the witnesses. In contrast, the other 22 representatives, including even those who are resistant to climate-change policy or critical of geoengineering, adhered to a standard of polite decorum and, in particular, treated the panelists with statements of respect, recognizing them as expert witnesses presenting on invitation from the Committee.

Ralph Hall, ranking Republican member from Texas, embodied this latter approach of collegial disagreement, in stark contrast to the style of Rohrabacher and Smith. His opening statement at the first hearing encapsulated this position, drawing upon the common narrative that suggests a sense of uncertainty and inconclusiveness:

> As many of my colleagues will agree, the debate about climate change is far from over, and I am sure that you have conducted and participated in that and came to the conclusion that the fact that there are still many, many opinions as to the causes, the effects and the potential solutions demonstrates how much uncertainty there is out there and how crucial it is for our Nation to continue to search for answers. (Hall, Congressional Hearing 2009a)

While Hall was explicitly critical of geoengineering and climate-change remediation, he spoke politely to the witnesses and Committee Chairman, along with some jovial jest with the latter.

Later in his opening statement at this first hearing on geoengineering, Hall tried to highlight the outlandishness of geoengineering proposals with an opaque analogy to Alfred Hitchcock's *The Birds* being adapted to "flying elephants," stating:

> So I think I have to thank you again, Mr. Chairman. This kind of opens up, you know—Alfred Hitchcock did *The Birds*. You remember that movie? And I have been working all since that time on a movie that have the elephants, flying elephants, you know, like Hitchcock had those birds that just were going to disturb the whole world. I don't know if I can get that underway or not, but we will maybe work that in in some of this here. (Hall, Congressional Hearing 2009a)

Hall then closed his opening remarks with an apparent joke, saying with a smile, "I would yield back to my Chairman, James Bond, and I thank you very much for letting me talk." Referring to Chairman Gordon as James Bond seemed to indicate that Hall considered the Chairman himself, with his interest in convening these hearings, to be immersed in science fiction.

Following Hall's remarks, Chairman Gordon said, "Well, Professor Shepherd [panelist from UK], welcome to America" and both the congressional members laughed at the interchange (Congressional Hearing 2009a). The banter between Chairman Bart Gordon and ranking member Ralph Hall seemed to counteract tension in the otherwise serious partisan posturing over controlling the parameters of debate. For example, Hall's use of self-deprecation regarding his lack of understanding geoengineering concepts provided a form of diffusion for tensions over substantive matters. At one point in the first hearing, Chairman Gordon was recommending the Royal Society Report as an informative source on geoengineering. He joked: "I was thinking about giving Mr. Hall the two-page summary, but I didn't want to overwhelm him . . ." to which Hall self-deprecatingly replied: "You would have had to read it to me" (Congressional Hearing 2009a). Humor between these two committee members was used to smooth over a significant ideological and policy gap in respect to the issue of climate change and, at these hearings, the possibility of geoengineering.

This humor, however, was intermixed with statements professing respect, which tempered the dynamic. On the occasion of the third hearing on geoengineering, Hall stated the following as his opening remarks:

> Thank you, Mr. Chairman, and but for my respect for you, I would have a lot longer opening remark here, but I would just say that I believe this is the third hearing our committee has held on geoengineering. As I have expressed on previous occasions, I have significant reservations about pursuing this line of research. With that, in the interest of time and courtesy to our very distinguished guest, I will just put this [written statement] in the record. (Congressional Hearing 2010b)

Hall's expression of "reservations" regarding the pursuit of geoengineering research referred to his position as a climate-change skeptic. In the same hearing, when questioning Hon. Phil Willis, a member of U.K. Parliament serving for the first time ever as a witness in a U.S. Congressional Science and Technology Committee hearing, Hall stated: "I am not terribly enthusiastic about this, but I am excited about your appearance here and the Chairman's vision" (Congressional Hearing 2010b).

Ralph Hall and Dana Rohrabacher represent two extremes of approach in addressing climate change and geoengineering from a climate skeptic perspective. What they have in common is that their position on climate change informs their position on geoengineering. Because climate change is not a problem by their estimation, pursuing geoengineering as a potential solution to climate change would be a waste of government resources in their view. This position was recurrent in the first three hearings on geoengineering, which occurred in 2009 and 2010. However, between the third hearing in March 2010 and the fourth hearing in November 2017, a shift occurred in which geoengineering became embraced by Republican politicians known as climate skeptics.

A SHIFT IN POLITICAL CONSIDERATION OF GEOENGINEERING: 2009–2017

During the course of the four hearings over eight years, with the first three clustered from November 2009 through March 2010 and the fourth in

November 2017, there have been some points of political consistency and some notable change over time. The level of committee members' understanding of geoengineering concepts increased over time. The narratives employed in partisan political performance, particularly among Republicans, changed over time, with a particular departure between the first three meetings and the fourth. Moreover, in that same time period, the composition of the committee itself transformed. These elements resulted in an evolution from the first hearing in November 2009 through the fourth hearing in November 2017, by which time a party-based stabilization emerged around particular positions on geoengineering.

In 2009, there was little understanding on the parts of many of the congressional representatives about the topic, as demonstrated through questioning and statements. The witnesses spoke most about SRM as a topic and much discussion of geoengineering was at a vague level. Only one committee member at the first hearing, Suzanne Kosmas (Democrat of Florida), took an explicit position in favor of pursuing geoengineering. By 2017, most of the congressional representatives seemed to have some understanding of the concepts. Witnesses and representatives spoke individually about each category, both SRM and CDR, in more specific terms. At the fourth hearing, eight out of 11 participating committee members indicated favorability toward pursuit or research of some form of geoengineering, with the details differing by party.

In terms of partisan-based political narratives and performances, the Democrats on the committee remained consistent in their positions on climate change, although they did stabilize in their position on geoengineering over time (as will be discussed). In contrast, the Republican narratives and discourse transformed remarkably between the first hearings and the fourth. In 2009, Republicans spent appreciable time disputing the existence of climate change and questioning the relevance of the topic of geoengineering given the "uncertainty" and "debate" about climate change. By 2017, several prominent climate-change deniers emerged as the most enthusiastic proponents of geoengineering. This was vividly portrayed in Randy Weber's (R-TX) declaration that "the future is bright for geoengineering!" (Congressional Hearing 2017b). In the fourth hearing, all the participating Republican members either avoided discussion of climate change or made active efforts to decouple geoengineering from climate change. Among this group of Republican politicians, the interest is in albedo modification (solar radiation management, SRM, is the term most used), with no mention of CDR, fitting with a worldview unconcerned with GHGs.

While only Republicans questioned the existence of climate change, political party was not determinative of individual Committee members' positions on climate change. During the first three hearings, four of the seven participating Republican representatives acknowledged the existence and importance of climate change. Only three of the seven challenged the existence of climate change. These three were vociferous on the issue, however, and two of the three participated in multiple hearings such that five of nine incidents of Republican participation included challenging the existence of climate change, serving to magnify that "side" of the "debate." Then at the fourth hearing, there was the shift in which the Republicans separated the issues of geoengineering and climate change. In contrast to the previous hearings, none of the six Republicans participating in the fourth hearing acknowledged climate change as an important issue to consider (or even debate) and several of them showed great enthusiasm for geoengineering technologies.

Figure 6.1 provides a visual representation of the participating Republican committee members' positions on climate change as expressed at the four geoengineering hearings. While the second and third hearings were more sparsely attended, it is clear that over the course of the first three hearings there is a split in the Republican approach to climate change as a subject. Comparing the first and fourth hearings, which included an equal number of Republican participants, reveals the stark difference over time, from one Republican out of six at the first hearing remaining ambiguous on the subject of climate change to six out of six at the fourth hearing.

Only Republican positions are presented in figure 6.1 because variance is only found among the Republican participants. One hundred percent of Democrat participants at the four hearings acknowledged the importance of climate change as an issue for governance. The change over time in the distribution of Republican positions contrasts with the Democrat members' consistent reiteration of the primacy of emissions abatement paired with willingness to consider, or at least learn more about, geoengineering proposals. As the discussion became increasingly specific over time, the Democrat

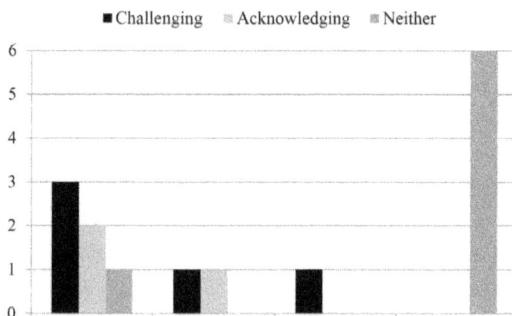

Figure 6.1   Republican Committee Members' Positions on Climate Change Based on Their Statements at the Four Geoengineering Hearings. *Source:* Author.

contingent stabilized around a position generally supportive of pursuing CDR and cautiously researching SRM, which was akin to the recommendations of the panelists.

At the first three hearings, the majority of Republican participants acknowledged climate change as an issue for governance. At the first hearing, one Republican, Brian Bilbray of California, articulated the politics around climate change in terms that instantiate Ingolfur Blühdorn's sociological concepts of "simulative politics" and the "politics of unsustainability" (Blühdorn 2000; 2011; 2007; Blühdorn and Welsh 2007) as discussed in earlier chapters. Representative Bilbray stated:

> I have come to the conclusion that we need to talk about mitigation of the crisis because we are not going to avoid it. There is not the political will to do what it takes. There is not even the political will to make it legal in the United States to do what it takes to avoid climate change because I believe strongly that we have got to have the ability to produce energy that doesn't emit greenhouse gases so we can shut down all those facilities that do, and there is not the political will to do with that what we did with the interstate freeway system where the government went out and sited, did the planning, did the things so we can shut down the coal producing and the emissions and all that other stuff. We are not willing to do that. We are just willing to talk about how terrible it is. (Bilbray, Congressional Hearing 2009a)

Bilbray's discussion of the lack of political will to "do what it takes" to address climate change through emissions abatement mirrors Blühdorn's language describing simulative politics: "despite their vociferous critique of *merely symbolic politics* and their declaratory resolve to take effective action, late-modern societies have neither the will nor the ability to *get serious*" (Blühdorn 2007, 253, emphasis in original). Bilbray reflexively acknowledges, verifies, and laments this critique of the political institution in which he serves.

In response to ongoing resistance from a subset of the Republican Party obstructing climate mitigation policy, Representative Baird, a Democrat from Washington, stated at the third hearing in March 2010:

> I think there is an urgent need for a constructive dialogue with my friends on the other side of the aisle on this, because we spend an inordinate amount of time here, on this committee, unfortunately, debating whether or not this is real, as if the outcome of our debate will somehow impact what happens in the real world. By that, I mean, as if climate change is going to be stopped if we declare it is not happening. But I think the adverse consequences that you [panelists] are describing, the profound geopolitical, national security, economic disruption

if you get your bet wrong, really has to be discussed. [. . .] Because if we just
say well, we are not going to do anything, because climate change is a hoax, as
is sometimes said by colleagues, that hoax can have some darn serious consequences if it is not a hoax. (Baird, Congressional Hearing 2010b)

As it turned out, however, in subsequent years, moderate Republicans supporting climate policy disappeared from the House Science and Technology
Committee. With the success of the "Tea Party" movement starting in
2009 and subsequent years along with other factors or political polarization, moderate Republicans were replaced by more ideologically extreme
Republicans in Congress. This led to the disappearance of moderate members such as Representative Bilbray quoted above regarding the lack of
political will, as well as Vernon Ehlers (R – Michigan) who took geoengineering seriously based upon its intrinsic premise of addressing climate
change. They were replaced by Republicans who took a partisan position
on climate change.

By 2013, climate-change denial became the majority view on the committee with attrition of moderate Republicans who "accepted climate science"
and "supported some climate action" (Lavelle and Hasemyer 2017). This
compositional shift of the committee is one component of the discursive
trends observed over the eight years of geoengineering hearings. Other components may include increasing awareness of climate change among politicians and their constituents at odds with their previous political investment
and entrenchment of climate-change denial, paired with a growing awareness,
understanding, and mainstreaming of geoengineering concepts and related
economic interests. In any case, once denialists claimed a dominant voice
within the Science, Space, and Technology Committee, geoengineering's
increased popularity among Congressional Republicans was paired with its
being rhetorically detached from climate change.

It might have been expected that the shift toward climate denial of the
Republican contingent on the committee would result in more discussion of
climate denial narratives as exhibited previously by Dana Rohrabacher and
Ralph Hall during the earlier hearings, but this was not the case. Rather, the
2017 geoengineering hearing was unique in the Republican contingent of the
committee generally using one of two approaches, either avoiding the topic of
climate change altogether or obfuscating anthropogenic climate change while
dismissing its relevance from the discussion of geoengineering.

Lamar Smith, chair of the House Science Committee, in an odd way came
the closest to acknowledging the relevance of climate change. He stated:

As the climate continues to change, geoengineering could become a tool to
curb resulting impacts. Instead of forcing unworkable and costly government

mandates on the American people, we should look to technology and innovation to lead the way to address climate change. Geoengineering should be considered when discussing technological advances to protect the environment and geoengineering should not be ignored before we have an opportunity to discover its potential. (Smith, Congressional Hearing 2017b)

This statement does refer to a changing climate in relation to the potential for geoengineering, but in a singular manner. It opens with the premise that "the climate continues to change," which, while it could be interpreted to be referring to climate change in the common meaning, leaves open the interpretation of the climate denial argument that the climate has always changed irrespective of people and GHG emissions. The second sentence, which criticizes government regulation of emissions, is the heart of Lamar Smith's position on climate change. It also gets the closest to acknowledging and engaging with the concept of anthropogenic climate change through its implied critique of emissions reductions as Plan A. Smith asserts that geoengineering might supplant the global community's Plan A. (This position is in stark contrast to the abundant caveats from geoengineering researchers, including the panelists at these hearings, that solar geoengineering could not replace emissions reductions and should not be considered as an alternative). In the final sentence of the selection, geoengineering becomes recast as a technology to "protect the environment" (Smith, Congressional Hearing 2017b). In this newly claimed narrative of protecting the environment, the words "climate change" disappear.

Going farther than Smith, the two most enthusiastic supporters of solar geoengineering at this fourth hearing were the two Republicans from Texas: Randy Weber and Lamar Smith, Chair of the Science, Space, and Technology Committee as of January 2013. Both of these Texan representatives have been vocal in climate-change denial, including within committee hearings (Mervis 2014), and have close ties to the oil and gas industries. For example, oil and gas interests have made up the largest aggregate contributors to Lamar Smith over his career, contributing over $772,000 to his campaigns from 1989 through 2018 (The Center for Responsive Politics 2018; Lavelle and Hasemyer 2017).[6] The prominence of these two members' pro-geoengineering narratives coexisting with climate denial reflects the changing composition of the committee. This compositional transformation accounts for some of the changes seen between the 2009/2010 and 2017 geoengineering hearings, including the support of some Republicans for climate policy in the early hearings which was absent in 2017. However, the decoupling of climate change from geoengineering and the avoidance of climate change from the Republican discourse during the fourth geoengineering hearing extends beyond these changes.

The result is significant newfound Republican support for geoengineering within Congress. While witnesses continued to stress the importance of mitigation and continued to advocate *de minimus* research, including climate models and cautious consideration of small-scale lab and field experiments, some Republicans indicated interest in actively pursuing SRM at a level significantly beyond witness recommendation. This new political support for geoengineering among Republican representatives parallels the broader trend of geoengineering moving from the margins of scientific discussion into mainstream consideration, but is unique in the element of discursively decoupling geoengineering from climate change.

In contrast to the more marked change in the discourse of Republican committee members, the positions of Democrat members remained fairly consistent, however with a gradual increase in favorability toward geoengineering research. Throughout the four hearings, Democrats reiterated the importance of mitigation and affirmed the existence of anthropogenic climate change. By the third hearing, the Democratic members indicated overall favorability to supporting geoengineering research. At the fourth hearing, Democratic members indicated favorability to pursuing CDR and researching SRM. The Democrats' positions on geoengineering research and the related framing and characterization of geoengineering tended to more closely parallel that of the witness panels, which largely emphasized the continued importance of emission reduction, discussed the role of CDR contributing to this, and provided caveats regarding risks of SRM, but nevertheless encouraged research into SRM as a potential consideration in terms of a "Plan B," as will be discussed.

Within his opening statement at the first hearing in November 2009, Chairman Gordon framed geoengineering and articulated the premise of hearings on the topic in the following manner:

> Geoengineering carries with it a tremendous range of uncertainties, ethical and political concerns, and the potential for catastrophic environmental side-effects. But we are faced with the stark reality that the climate is changing, and the onset of impacts may outpace the world's political and economic ability to avoid them. Therefore, we should accept the possibility that certain climate engineering proposals may merit consideration and, as a starting point, review research and development as appropriate. At its best geoengineering might only buy us some time. But if we want to know the answers, we have to begin to ask the tough questions. Today we begin what I believe will be a long conversation. (Gordon, Congressional Hearing 2009a)

This framing of geoengineering is reminiscent of content and tone found in the Royal Society report (2009) as well as the later National Academy report (National Research Council). It draws upon the conventionalized narratives of geoengineering as a tool in case of catastrophe, as a stopgap, as a Plan B, and also the distinction between research and deployment developed within science policy reports.

Later in the hearing, Chairman Gordon referred to the Royal Society report in its capacity, at the time, as geoengineering's primary document for those trying to understand the concepts. Gordon stated: "I would really advise that anyone that has an interest in this issue to review the Royal Society's report. It is very good" (Gordon, Congressional Hearing 2009a). This speaks to the significance of science policy reports in influencing policy makers and their understandings of scientific and technological concepts. Chairman Gordon's opening remarks, clearly influenced by his reading of the Royal Society report, are also exemplary of the trend seen throughout the four hearings of Democrats largely matching their discourse on geoengineering with the terms, narratives and framings of scientific experts, including those serving on the witness panels, some of whom were also involved in the writing of the key science policy reports on geoengineering.

The following discussion will specifically examine the role of scientific witnesses in the context of these hearings.

WITNESS TESTIMONY

Panelists who participated as expert witnesses at the four geoengineering hearings were mostly scientists involved with geoengineering research. As mentioned, each panel was comprised of two to four witnesses from academic institutions in addition to one or two affiliated with either a national laboratory or a think tank.[7] The March 2010 hearing also included one representative from GAO and MP Phil Willis. See table 6.2 for a full listing of panelists from all four geoengineering hearings.

While a shift can be observed in the treatment of geoengineering by members of Congress, the panels of expert witnesses remain, for the most part, consistent in their discussions of geoengineering over time. At the fourth hearing, one witness, Douglas MacMartin, made the statement: "I think one of the striking things about this panel is actually how broad our agreement is likely to be on almost all of the issues" (Congressional Hearing 2017b). This assertion holds generally true across the 18 instances of witness testimony over four hearings,[8] with a few exceptions. Overall, the panelists showed broad agreement in terms of both advocating for geoengineering research and also expressing the importance of mitigation irrespective of geoengineering. All but two witnesses articulated the importance of traditional mitigation in their written and/or oral testimony and all but three placed strong emphasis on the importance of emission abatement irrespective of geoengineering pursuit.

Table 6.2 Witnesses to Congressional Geoengineering Hearings

Witness	Position	Affiliation (at time)
Geoengineering: Assessing the Implications of Large-Scale Climate Intervention, November 5, 2009		
Dr. Ken Caldeira	Professor of Environmental Science	The Carnegie Institution of Washington
Professor John Shepherd	Professional Research Fellow in Earth System Science	National Oceanography Centre
Mr. Lee Lane	Codirector, American Enterprise Institute Geoengineering Project	American Enterprise Institute (AEI)
Dr. Alan Robock	Professor, Department of Environmental Sciences	Rutgers University
Dr. James Fleming	Professor and Director, Science, Technology and Society Program	Colby College
Geoengineering II: The Scientific Basis and Engineering Challenges, February 4, 2010		
Dr. David Keith	Research Chair in Energy and the Environment	University of Calgary
Dr. Philip Rasch	Climate Scientist	Pacific Northwest National Laboratory
Dr. Klaus Lackner	Geophysicist, Earth and Environmental Engineering	Columbia University
Dr. Robert Jackson	Chair of Global Environmental Change	Duke University

Geoengineering III: Domestic and International Research Governance, March 18, 2010

Hon. Phil Willis, MP	Chairman, Science and Technology Committee	United Kingdom House of Commons
Dr. Frank Rusco	Director of Natural Resources and Environment	Government Accountability Office (GAO)
Dr. Granger Morgan	Professor, Department of Engineering and Public Policy	Carnegie Mellon University
Dr. Jane Long	Geotechnical engineer	Lawrence Livermore National Laboratory
Dr. Scott Barrett	Professor of Natural Resource Economics	Columbia University

Geoengineering: Innovation, Research, and Technology, November 8, 2017

Dr. Phil Rasch	Climate Scientist	Pacific Northwest National Lab
Dr. Joseph Majkut	Director of Climate Policy	Niskanen Center
Dr. Douglas MacMartin	Senior Research Associate	Cornell University
Ms. Kelly Wanser	Principal Director, Marine Cloud Brightening Project	University of Washington

The two most notable exceptions were Lee Lane of the conservative think tank, American Enterprise Institute, and geophysicist Klaus Lackner of Columbia University. Lee Lane promoted pursuit of SRM as a top research priority and generally avoided the topic of emissions abatement. However, while answering a question he did concede that he did not "believe that we can go on emitting greenhouse gases at ever-increasing rates" and "eventually controls are going to be essential," but that "conditions are not in place yet" (Congressional Hearing 2009a). Lackner emphasized the importance of achieving a "carbon neutral energy economy" (Congressional Hearing 2010a). His testimony was unique, however, in his expressed belief that CDR could be relied upon to achieve carbon neutrality without necessarily converting from a carbon-intensive energy economy. Kelly Wanser, the Principal Director of the Marine Cloud Brightening Project, speaking at the fourth hearing, displayed a position between these outliers and the otherwise consistent reiteration of the importance of mitigation among the other 14 unique witnesses. Wanser mentioned mitigation as one component that should be used along with other strategies in a portfolio of responses to climate change, but she did not emphasize the importance of mitigation to the same extent as did most witnesses (Congressional Hearing 2017b).

Apart from these three exceptions, the other witnesses were all unwavering in their emphasis on the primacy of mitigation. Exemplifying this position, Ken Caldeira stated simply: "Climate change poses a real risk to Americans. The surest way to reduce this risk is to reduce emissions of greenhouse gases, such as carbon dioxide" (Congressional Hearing 2009a). Alan Robock in his testimony made explicit his agreement with Caldeira, stating:

First I would like to agree with Ken Caldeira, that global warming is a serious problem and that mitigation, reduction of emissions, should be our primary response. We also need to do adaptation and learn to live with some of the climate change which is going to happen no matter what. (Congressional Hearing 2009a)

Robert Jackson opens his testimony on geoengineering by stating the importance of emission reductions:

Let me begin by stating that a wealth of evidence already shows our climate is changing and is a threat to people and organisms. As a scientist and citizen of our great Nation, I urge you to act quickly to reduce greenhouse gas emissions. (Congressional Hearing 2010a)

Moreover, witnesses emphasized the primacy of mitigation *irrespective* of geoengineering. Robert Jackson's written testimony expanded on his call to emissions reductions with the caveat: "The safest, cheapest, and most prudent way to slow climate change is to reduce greenhouse-gas emissions soon. No approach—geoengineering or otherwise—should lead us from that path" (Jackson written statement, Congressional Hearing 2010a). Alan Robock stated:

> I would just like to say that we can't hold geoengineering as a solution and allow that to reduce our push toward mitigation. It is never going to be a complete solution. We may need it in the event of an emergency, but let us not stop mitigation and wait and see if geoengineering would work. That is not the right strategy. (Congressional Hearing 2009a)

Summarizing many of these shared sentiments, John Shepherd, the lead author of the Royal Society Report, stated:

> Since time is short, I would like to move directly to summarize the key messages of our study and first among these is that geoengineering is not a magic bullet. None of the methods that have been proposed provide an easy or immediate solution to the problems of climate change. There is a great deal of uncertainty about various aspects of virtually all the schemes that are being discussed. So at present, this technology, in whatever form it takes, is not an alternative to emissions reductions which remain the safest and most predictable method of moderating climate change, and in our view cutting global emissions of greenhouse gases must remain our highest priority. (Shepherd, Congressional Hearing 2009a)

Through these various statements and more, the panelists clearly expressed that geoengineering is not a replacement for mitigation.[9]

Even within the context of advocating for the importance and potential of geoengineering, the scientists carefully noted the limitations of geoengineering and reiterated the primacy of traditional mitigation. For example, immediately after expressing the promising potential of SRM, Ken Caldeira candidly states at the first hearing: "Nobody thinks these approaches will perfectly offset the effects of carbon dioxide" (Congressional Hearing 2009a). Similarly, at the fourth hearing, Douglas MacMartin articulated: "It is important to stress at the outset that solar geoengineering cannot be a substitute for cutting emissions for several reasons. This conclusion has been reached by every assessment of this technology, including by the National Academies in 2015" (Congressional Hearing 2017b).

David Keith, as discussed in the previous chapter, is one of the most prolific advocates of SRM pursuit with the bold position that SRM should not necessarily be reserved as a distant Plan B (e.g., see Keith 2013, 172–173;

Keith 2016). He insisted, however, that this position on solar geoengineering in no way diminishes his emphasis on the importance and urgency of emissions reductions. He stated: "We must make deep cuts in global emissions if we are going to manage the risks of climate change. Emissions reductions are necessary, but they are not necessarily sufficient" (Keith, Congressional Hearing 2010a). Keith's written statement emphasized that geoengineering and emissions abatement are not mutually exclusive: "Responsible management of climate risks requires sharp emissions cuts and clear-eyed research and assessment of SRM capability. The two are not in opposition. We are currently doing neither; action is urgently needed on both" (Keith written statement, Congressional Hearing 2010a). Similarly, Granger Morgan, a proponent of SRM pursuit,[10] included in his testimony a statement consistent with the panelists' collective emphasis on mitigation:

> I want to emphasize that I am not arguing that the U.S. or anybody else should engage in SRM. The U.S. and other large emitting countries need to get much more serious about reducing emissions and lowering the concentration of atmospheric carbon dioxide. I believe that can be done at an affordable cost. (Morgan, Congressional Hearing 2010b)

In addition to the emphasis on the importance of addressing emissions irrespective of geoengineering, the witnesses also engaged with several other narratives common to geoengineering discourse, including the idea of geoengineering as a "last resort" or "Plan B" and the related framing of geoengineering as a potential tool to address a climate catastrophe or emergency. Numerous scientists characterized geoengineering as "a last resort." Philip Rasch wrote: "Geoengineering should be viewed as a choice of last resort[.] It is much safer for the planet to reduce greenhouse gas emissions. Geoengineering would be a gamble" (Rasch written statement, Congressional Hearing 2010a). Jackson closed his testimony saying: "In conclusion, although emitting less carbon dioxide and other greenhouse gases should remain our first priority [. . .] We need to get geoengineering right as a tool of last resort" (Congressional Hearing 2010a). Phil Willis stated: "this is an issue of last resort and must not, in fact, deflect us from our major task of making sure that we put less CO_2 into the air, and where it is there, that we look, in fact, to sequestrate it" (Congressional Hearing 2010b).

Of the 18 witness instances, 13 employed catastrophe framing as is common in regard to geoengineering. Catastrophe framing is a variant of "Plan B" framing in which geoengineering is treated as a tool that needs to be available in case of a climate emergency (cf. Nerlich and Jaspal 2012; Corner, Parkhill and Pidgeon 2011, 13). Alan Robock, who, as discussed previously, is known for being particularly candid in his concerns regarding geoengineering, stated at the first hearing on the subject: "Using geoengineering should only be in the event of a planetary emergency and only for a temporary period of time, and it is not a solution to global warming" (Congressional Hearing 2009a). In this way, Robock engaged the common stopgap and climate emergency framings but used these framings as a caveat, not an imperative. Also at the first hearing, John Shepherd, lead author of the Royal Society report, stated: "in our view, this is not a technology which is ready for deployment in the immediate future. It is, however, a technology that may be useful at some point in the future if we find that we have need of it" (Congressional Hearing 2009a). In addition to this implicit use of catastrophe, or climate emergency, framing, Shepherd later refers to geoengineering as a Plan B: "What we need is research on a small portfolio of promising techniques of both major types in order that our Plan B will be well prepared, should we ever need it" (Congressional Hearing 2009a).

For the third hearing, Jane Long of Lawrence Livermore Laboratory presented the catastrophe framing premise within her written statement:

> In this future, if climate sensitivity (the magnitude of temperature change resulting from a doubling of CO_2 concentrations in the atmosphere) turns out to be larger than we hope or mitigation proceeds too slowly, we cannot rule out the possibility that climate change will come upon us faster and harder than we—or the ecosystems we depend on—can manage. (Long written statement, Congressional Hearing 2010b)

Phil Willis, the U.K. minister who spoke at the third hearing, provided another "Plan B" framing closely related to the catastrophe framing given by Robock, Long, and others: "If the climate warms dangerously, and we can't fix the problem by reducing carbon emissions or adapting to the changing climate, geoengineering might be our only chance" (Willis, Congressional Hearing 2010b). This framing of geoengineering as being in reserve in case of a climate emergency is consistent with that in the major science policy reports, as discussed previously.

As may be expected due to the selection criteria for expert witnesses, all panelists promoted the need for geoengineering research with one possible exception. The overall promotion for research was consistent with the policy recommendations of science policy reports including the Royal Society's report and the NRC's report. Individual witnesses provided their own visions of how research might be best promoted and coordinated, but every witness promoted research. Ken Caldeira aptly articulated at the first geoengineering hearing: "while the panel disagrees about maybe the scale and scope of what a research program should be, I think it is indicative that the entire panel asserts the need for a research program" (Congressional Hearing 2009a). This observation generally holds throughout the four panels.

The one exception was science historian James Fleming, who was the lone social scientist included among the witness panels. He encouraged research on human dimensions related to the pursuit of geoengineering, but did not take an explicit position on geoengineering research itself, presumably as it falls outside his area of expertise (Congressional Hearing 2009a). The historian did, however, include a number of past examples of human hubris in relation to trying to control weather or climate, as well as the fallacies and risks involved. He cautioned learning from history's lessons. While not offering specific recommendations, Fleming indicated that mitigation is necessary and displayed reluctance regarding geoengineering, implying it may be a foolhardy endeavor. Fleming ended his testimony by saying:

> People have said that climate control is not a good idea. Harry Wexler, head of research at the Weather Bureau, said this in 1962, and just two years ago, Bert Bolin, the first chair of the IPCC, wrote that the political implications of geoengineering are largely impossible to assess and it is not a viable solution because in most cases, it is an illusion to assume that all possible changes can be foreseen. Climate change is simple. We should do the right thing. Climate is complex. It involves oceans and atmospheres, ice sheets and now monsoons, so studying the human dimension is essential. We need the interdisciplinary, international and intergenerational emphasis. (Fleming, Congressional Hearing 2009a)

In context, the otherwise vague statement that "We should do the right thing" implies mitigation through emission reductions and "Climate is complex" is a warning in regard to hubristic ventures attempting climate control (Fleming, Congressional Hearing 2009a). However, he does not explicate an opposition to technical research, only gives the implicit warning and indicates that any research that does occur should include "interdisciplinary, international and intergenerational emphasis" (Fleming, Congressional Hearing 2009a).

The other 17 instances of witness testimony included some degree of promoting geoengineering research, within a broad range of opinions of what research should be pursued and in what ways. For instance, Alan Robock, who consistently voices cautions in regard to SRM, specified that in terms of albedo modification, the only research he recommends is climate models and no field tests. He stated: "I would like to urge you to support a research program into the climatic response with climate models, into the technology to see if it is possible to develop different systems so that you can make an informed decision in the future" (Robock, Congressional Hearing 2009a).

While encouraging research in the form of climate models, he was very clear in his position against field testing, arguing:

> If we wanted to do experimentation, it is not possible to do just a small-scale test [. . .] so we would really have to put a lot of material in for a substantial period of time to see whether we are having an effect. And that would essentially be doing geoengineering itself. You can't do it on a small scale [. . .] it is problematic whether we could actually ever do an experiment in the stratosphere without actually doing geoengineering. (Robock, Congressional Hearing 2009a)

In contrast, during the same hearing, Ken Caldeira indicated that there are "small-scale field studies that could be done short of something that affects climate" (Congressional Hearing 2009a). Along these lines, at the second hearing, David Keith argued in his written statement: "Field tests will be needed, such as experiments generating and tracking stratospheric aerosols to block sunlight and dispersing sea-salt aerosols to brighten marine clouds" (Keith, written statement, Congressional Hearing 2010a). Keith further averred:

> Although risk of climate emergencies may motivate SRM research, it would be reckless to conduct the first large-scale SRM tests in an emergency. Instead, experiments should expand gradually to scales big enough to produce barely detectable climate effects and reveal unexpected problems, yet small enough to limit resultant risks. (Keith, written statement, Congressional Hearing 2010a)

In this way, Keith embraced the climate emergency framing, but recommended a gradual and preemptive implementation of geoengineering.

Philip Rasch took a position somewhere between that of Robock and that of Caldeira and Keith. Like Caldeira and Keith, Rasch indicated that his vision of a research program would include a fieldwork component:

> Lab and fieldwork are critical to assure a thorough understanding of the fundamental physical process important to climate and that computer models are reasonably accurate in representing that process. I think it is critical to distinguish between "small scale field studies" where we might introduce some particles into the atmosphere over such a small scale that they would have negligible climate impact, and "full scale deployment" where we expect to actually have a climate impact. *Field studies might try to induce a deliberate change to some feature of the earth system at a level with a negligible impact on the climate, but the change would allow us to detect a response in a component important to climate.* (Rasch, written statement, Congressional Hearing 2010a, 154, italics in original)

However, like Robock, Rasch problematized the distinction between research and deployment and the difficulty of achieving meaningful field results without crossing over the line toward deployment. Rasch noted that there "will be substantial difficulties in evaluating this geoengineering strategy without full deployment. This makes it difficult to improve our understanding slowly and carefully using field experiments that do not change the Earth's climate" (Rasch, written statement, Congressional Hearing 2010a, 157). He went on to add nuance to this argument in terms of the challenges involved with field tests, concluding with an articulation of the resulting conundrum:

> So we are caught between rock and a hard place. Too small a field test, and it won't reveal all the subtleties of the way the aerosols will behave at full deployment. A bigger field test to identify the way the aerosols will behave when they are concentrated will have an effect on the planet's climate (like Pinatubo did), albeit for only a year or two. I have not seen a suggestion on how to avoid this issue. (Rasch, written statement, Congressional Hearing 2010a, 157)

Rasch's statement on this matter exemplifies the internal tension regarding recommending field tests but with extensive caveats regarding the thin, even elusive, line between research and deployment.

Granger Morgan was more sanguine regarding the possibility of drawing lines to differentiate various levels of experimentation. In fact, he proposed doing just that. He included in his testimony at the third hearing, diagrams that conceptualize the distinction of experimentation of a *de minimus* standard that he proposed should be exempt from additional oversight and governance apart from "transparent public announcement and informal coordination within the scientific community" (Morgan, slides, Congressional Hearing 2010b). For example, one of his slides shows a diagram in which

> X, Y, and Z define the limits of an allowed zone. They refer, respectively, to the upper bounds on the amount of radiative forcing that an experiment might impose, the duration of that forcing, and the possible impacts on ozone depletion. (Morgan, Congressional Hearing 2010b)

However, the values of X, Y, and Z, as well as "what forms of international agreement and enforcement, if any, would be most appropriate, and what scientific input would they require" was left as an open question (Morgan, Congressional Hearing 2010b).

These various positions on the pursuit of field experimentation underscore the variety of opinions held within the community of geoengineering researchers. It speaks to a tension in regard to the potential for knowledge-building through field tests, the risks that would be involved with field tests, and the nebulous border between experimentation and deployment. This tension can be seen embodied in an individual scientist's reasoning, as demonstrated in Rasch's statement cited above, or in diversity of competing opinions on this topic among the members of geoengineering research community, the so-called geoclique (see discussion in chapter 4 and Hamilton 2013a; Kintisch 2010). STS scholar Nils Markusson argues "that the geoengineering imaginary is [. . .] about the creation of a new scientific space for the conversion of climate science into applied, experimental technology, and that the boundaries and the very desirability of this space are contested" (Markusson 2013, 3). The points of consensus and the points of tension demonstrated by the panelists within these hearings provide a window into the variously solidifying and contested boundaries of the space.

Markusson's (2013) research notes the ambivalence that manifests in geoengineering reports authored by a collective process. The variance between the witnesses giving testimony at these hearings speaks to this issue of diversity and the resulting ambivalence in some of the co-authored articulations of the geoengineering imaginary. The data being considered here complements Markusson's (2013) findings, since the witnesses at these hearings are exemplary of the cross-section of authors participating in collectively authored geoengineering reports. Notably, of the 17 unique individuals who gave testimony at the congressional hearings, seven were involved in the writing or review of either the 2009 Royal Society report or the 2015 National Research Council report, the two most important geoengineering science policy reports (discussed at length in Section II). The examples discussed above regarding the range of opinions on field experimentation, were all drawn from this subset of seven witnesses who were also involved in the science policy reports. Hence, these individual articulations of the geoengineering imaginary and the envisioning of what a geoengineering research program would include provide a nuanced perspective into the diverse range of positions that become encapsulated within influential geoengineering reports.

Other points of disagreement emerge in terms of the consideration of deployment and the economic arguments drawn upon by advocates to promote solar geoengineering as a relatively inexpensive venture. For example, at the first geoengineering hearing, Alan Robock, one of the most precautionary geoengineering researchers, raised doubts or questions to some of the statements of other panelists. For instance, in his testimony, Ken Caldeira had drawn upon a narrative common in geoengineering discourse, asserting: "Preliminary research suggests that we could rapidly and relatively cheaply put tiny particles high in the stratosphere and that this would cause the earth to cool quickly" (Congressional Hearing 2009a, 16). Then Lee Lane of AEI pushed this narrative further, making the argument in his testimony that SRM would be a low-cost investment with high potential benefit (Congressional Hearing 2009a, 33).

During his testimony, Robock raised questions regarding the assumed cheapness and ease of execution of SRM: "Ken [Caldeira] said it would be easy and cheap, but there is no demonstration of that. It might not be that expensive, but such equipment just doesn't exist today" (Congressional Hearing 2009a, 44). He also critiqued the economic analysis cited by Lee Lane, saying: "I disagree with the economic analysis because they just ignored many of the risks and didn't even count what the possible dangers might be" (Robock, Congressional Hearing 2009a, 45). However, Robock was in agreement with the central argument of advocating research. He continued:

> But I agree with everybody that we need a research program so that we can quantify each of these [risks and dangers] so policymakers can tell if—is there a Plan B in your pocket, or is it empty? We really need to know that, and we don't know the answer to that yet. (Robock, Congressional Hearing 2009a, 45)

While the promotion of research is entirely consistent with the other panelists, Robock's framing of the Plan B narrative was more cautious than its typical use. He advocated research to establish *whether* there is a Plan B as opposed to preparing Plan B for readiness.

Moreover, Robock was the most consistent witness in returning the discussion to the issue of risks, as opposed to simply mentioning risks and then pivoting toward other issues of interest. For example, as discussed in chapter 2, volcanoes are frequently used as an analogue for stratospheric aerosol albedo modification. This analogue was brought up a number of times at the hearings both in oral statements and in the written record. Ken Caldeira, for instance, spoke in positive terms about how the idea of SRM is demonstrated through the natural experiment of large volcanic eruptions, explaining that SRM methods "seek to reduce the amount of climate change by reflecting some of the sun's warming rays back to space. We know this basically works because volcanoes have cooled the earth in this way" (Congressional Hearing 2009a, 16). Several other witnesses employed the volcanic analogy in a similar manner.

Later in the hearing, during the question period, Chairman Gordon had asked an unrelated question, but before answering the question, Robock returned to the volcano analogy to express the reverse side of the natural experiment:

> First of all, I would like to mention that although the Pinatubo volcanic eruption cooled the planet, it also produced drought in Asia and Africa. It destroyed

ozone, and it reduced solar radiation generation from direct solar radiation by 30 percent in those technologies that were developing. So it is a lesson of efficacy but also of problems. (Congressional Hearing 2009a)

After making this statement, he returned to the question asked. Robock's consistent role in reiterating the level of risk and complexity involved in solar geoengineering highlights the ease at which these novel technologies can become normalized in common discourse on geoengineering in the absence of someone so vigilantly returning to the risks. This is despite the fact that the far-reaching nature of geoengineering risks are broadly agreed upon within the geoengineering research community. However, as discussed elsewhere, risks are often mentioned before discussion is pivoted toward technical questions or else treated as footnotes, sometimes quite literally, as discussed in chapter 2 regarding the volcano analogy's treatment in the NRC report.

These examples highlight variance of opinions between panelists on topics such as field experimentation, the estimated costs of SRM, and degree of confidence in the feasibility of a Plan B. Despite the differing positions on these topics as well as on the specific details regarding how to best fund and coordinate research, the panels displayed consistent promotion of the importance of geoengineering research for both SRM and CDR. As Ken Caldeira noted, variance among witnesses on envisioning the details of a research program, highlights the consistency of the shared sentiment that there should be some form of research program (Congressional Hearing 2009a). The need to support research in some form or another received unanimous consensus among witnesses speaking at the four geoengineering hearings. Furthermore, with the exception of the few anomalies discussed earlier in this section, there was broad agreement on the urgency and primacy of mitigation and the related point that geoengineering cannot replace mitigation.

SITUATING GEOENGINEERING
HEARINGS IN CONTEXT

Congressional hearings in which committees hear from panels of invited experts officially serve several purposes including information-gathering either in regard to specific legislation or in an exploratory capacity. In addition, these hearings provide a platform for reenacting recurrent political arguments of differing levels of relevance to a topic. For example, within the geoengineering hearings, there is spillover of the political contention regarding climate change, which has also received its own hearings within the same Committee on Science, Space, and Technology. This includes Democrats (and, in the past, some Republicans) reiterating the existence of anthropogenic climate change and the urgency of mitigation and adaptation while a vocal contingent of Republicans reiterate the "uncertainty" regarding climate change and imply the science is still a matter of "debate." The hearings further included Republican recitations of narratives regarding economic competition, national debt, and economic responsibility, which influenced their engagement with climate change and geoengineering, based on ideology that prioritizes economic over environmental concerns. These themes can be seen in the geoengineering hearings and also other related hearings of the committee as well as structured into the hearings themselves through the selection of witnesses.

As mentioned in previous chapters, Aaron McCright and Riley Dunlap's (2010) research points to the ways in which the American conservative movement has managed the boundaries of climate policy. The climate skepticism movement that is related to this continues to cultivate the sense that there is an ongoing debate about the existence of anthropogenic climate change. The congressional hearings on climate change from recent years, especially since the committee transitioned to being majority skeptic, reaffirm this trend as demonstrated by the witness lists that have been designed to simulate debate within the formal structure of congressional hearings that are ostensibly meant to provide information to guide policy making. For example on March 27, 2017, the House Committee on Science, Space, and Technology held a hearing entitled "Climate Science: Assumptions, Policy Implications, and the Scientific Method," which included four panelists: three (Judith Curry, John Christy, and Roger Pielke Jr.) who dispute scientific consensus on climate change and one (Michael Mann) to represent the majority view on climate change (Congressional Hearing 2017a). Similarly, an earlier hearing of the Subcommittee on Environment on March 6, 2013 called "Policy Relevant Climate Issues in Context" included three panelists: two who have been vocal in disputing climate-change consensus (Judith Curry and Bjørn Lomborg, the self-labeled "skeptical environmentalist" discussed previously) and one (William Chameides) who has been vocal on the dangers and risks of climate change (Congressional Hearing 2013). In these instances, the Committee on Science and the Subcommittee on Environment designed the climate-change hearings to provide a performance of debate on climate change, a literal manifestation of the political narrative that contends that climate change is still up for debate. The witness panels of these climate-change hearings, moreover, were constituted by a majority of panelists presenting the minority views disputing climate change and only one scientist at each hearing to present the mainstream scientific perspective.

The geoengineering hearings provide an additional venue for performance of political posturing, presentation, and interaction with climate change and proposed policy approaches. Within the geoengineering hearings, a subset of Republican representatives reassert the narrative that the science of climate change is not settled, but rather still subject to an active debate. However, geoengineering complexifies the dynamic, by offering a "solution" that is acceptable to conservative interests, including this particular subset of Republican representatives, even while they continue to deny the relevance of the problem it addresses. This became evident at the fourth of the geoengineering hearings when Texan Republican representatives, known for positions against climate mitigation policy, came to enthusiastically embrace solar geoengineering as a favorable alternative to mitigation, or "regulation" as Chairman Smith refers to it (Congressional Hearing 2017b). Projecting albedo modification as an alternative to mitigation was in clear contrast to the core message from the four sets of panelists with few exceptions: that geoengineering should be researched, but that it is not a magic bullet, there is a primary need to reduce emissions, and that geoengineering is not an alternative to emissions reductions (Congressional Hearing 2009a; Congressional Hearing 2010a; Congressional Hearing 2010b; Congressional Hearing 2017b).

Members of both political parties recurrently reiterated their party's position on climate change and then pivoted the discussion in such a way that by the final two hearings on geoengineering, a majority of members of both parties ultimately express some level of support for pursuing geoengineering research and some degree of possible implementation. This is seen among the Democrat members' consistent reiteration of the urgency and primacy of mitigation and adaptation, but with discursive flexibility that leaves room for the consideration of geoengineering as an additional strategy, especially CDR. As discussed, at the fourth hearing, Republican committee members employed the discursive technique of decoupling climate change and geoengineering to facilitate the simultaneous support of otherwise contradictory positions, dismissing climate change and embracing solar geoengineering. Ultimately, while Democrats converged around a position very closely resembling the recommendations of the expert panelists, Republicans at the fourth hearing promoted the possibilities of albedo modification far beyond what the witnesses themselves recommended.

At the earlier hearings, members from both parties expressed concern about the lack of political will to match the urgency of climate change. Eight years later, climate policy remains an issue of speculation and the arguments that geoengineering may be necessary should mitigation efforts be insufficient have become increasingly pertinent. However, geoengineering would involve its own social and political challenges and risks in addition to environmental risks. Jane Long's written testimony for the third geoengineering hearing spoke directly to this. She wrote:

As we consider geoengineering, we have to recognize that society has not been able to quickly or easily respond to the climate change challenge. Consequently, the geoengineering option isn't just a matter of developing new science and technologies. It is also a matter of developing new social and political capacities and skills. (Jane Long, written statement, Congressional Hearing 2010b)

Yet, geoengineering, bringing new social-political challenges, cannot replace mitigation and its own web of long-standing social-political challenges. Geoengineering, then, presents an additional layer of political performance and maneuvering, a forum for the "politics of representation" (Mehan 2000; Mehan and Wills 1988) within an arena characterized by the "politics of unsustainability" (Blühdorn 2011; Blühdorn and Welsh 2007). While geoengineering cannot replace mitigation and comes with its own environmental, social, and political challenges, it provides another avenue for continued "debate" that further facilitates prolongating the status quo as climate mitigation policy decisions are delayed. Meanwhile, the stakes and risks involved continue to escalate, reshaping the environmental and political landscape in which eventual decisions will be made.

NOTES

1. Video recordings of the four hearings were methodically reviewed along with relevant written materials. The hearing videos, uploaded by Congress, are available for viewing through YouTube. (Last accessed April 7, 2018.) The official record for the first three hearings includes transcripts of hearing testimony as well as written statements and other documents submitted for the public record. At the time of the research, the written record for the November 2017 geoengineering hearing had not been released, so transcription was made directly from the video footage. Some written materials were available, such as the hearing's charter and written statements by the witnesses. Page numbers are included in citations of quotes drawn from the written material of the official record, including transcripts of hearing testimony, but not for quotations transcribed directly from the video recordings. The research process included close readings of the written materials and viewings of the hearing videos, transcribing select sections of the fourth hearing, memoing for all hearings, tracking themes, topics and actors within written materials using Nvivo, and creating spreadsheets to track data of relevant attributes and positions stated for all witnesses and participating committee members at the four hearings.

2. For instance, the first three hearings occurred during a critical period related to the major congressional issue of the time, healthcare reform. The first House Committee on Science and Technology hearing on the topic of geoengineering occurred the week that the House was debating and voting on the Affordable Care Act

(ACA). The third hearing occurred the week that Congress was voting to reconcile the House and Senate versions of the ACA.

3. H.R. 4586 was introduced in the House of Representatives on December 7, 2017 and referred to the House Committee on Science, Space, and Technology and on May 22, 2018 was referred to the Subcommittee on the Environment (McNerney 2017b).

4. H.R. 5519 was introduced in the House of Representatives on December 19, 2019 and referred to the House Committee on Science, Space, and Technology, which referred it to the Subcommittee on Environment, where it received consideration and mark-up and was forwarded by the Subcommittee back to Full Committee on March 4, 2020 (McNerney 2019).

5. The problematic discrepancy between "theory" as used in academic communities versus other general usages has increased in intervening years since this hearing. It is notable that the rise of "conspiracy theories" being discussed and contended in public discourse further confuses the public understanding of "theories" in the scientific sense. Scientific theories are based on logical interpretation and extrapolation of empirical fact. This, of course, is completely the opposite of so-called "conspiracy theories" that have no basis in fact or logic but attempt only to peddle falsehoods.

6. Incidentally, the fossil fuel industry has also been the greatest contributor to members of the House Science Committee overall since 2006 (Lavelle, 2017).

7. The two think tanks represented at these hearings were the conservative American Enterprise Institute (AEI) and the Niskanen Center, often characterized in terms of libertarian values.

8. Philip Rasch, a climate scientist at Pacific Northwest National Laboratory, served as a witness at two separate hearings (the second and the fourth). In total, there were 17 unique witnesses, and 18 instances of witness testimony, including Rasch's two appearances.

9. This, of course, is consistent with the science policy reports to which some of these panelists have contributed.

10. Recall that M. Granger Morgan was a co-author of the *Foreign Affairs* article discussed in the previous chapters.

Section V

TECHNOLOGY AND REFLEXIVITY

Conclusion

Interconnections, High Technology, and Reflexive Modernization

INTERACTIONAL EFFECTS, INTERCONNECTIONS, AND COMPLEXITIES

Geoengineering is controversial for a number of reasons, including its novelty, the magnitude of its risk, the challenges of governance involved, the regional disparity of potential effects, and concerns regarding moral hazard, meaning that its very consideration could impede emission reduction efforts (e.g., Hulme 2014; Szerszynski et al. 2013; Stilgoe 2016; Fischer 2017, 84–85). The implementation of geoengineering projects would bring about new risks and uncertainties inherent in the pursuit of such real-world experiments (e.g., Huesemann and Huesemann 2011; Beck 2009; Harris 2013; Hamilton 2013a; Parkinson 2010; Macnaghten and Szerszynski 2013; Szerszynski et al. 2013; Stilgoe 2016). Nevertheless, geoengineering is being mainstreamed as a topic of consideration within scientific communities, politics, and the public.

The various genres of discourse addressed here all contribute to and reinforce each other on reflecting and advancing the mainstreaming of geo-engineering. Scientific discourse, including articles, research disclosures, and science policy reports, especially the reports of prestigious scientific academies, represent a critical juncture in the trajectory of geoengineering. They are used as evidence by journalists of the mainstreaming of geoengi-neering within the scientific community and they have also been relied upon as a premise for government hearings and political consideration. In this sense, the very existence of the reports furthers mainstreaming. Moreover, discursive conventions within them help to normalize novel geoengineer-ing proposals and legitimize geoengineering research. News media, in the process of reporting upon the developments related to geoengineering, have reinforced the narrative of its move from "fringe" to mainstream and have themselves brought increased public attention to geoengineering since 2006. Congressional hearings held between 2009 and 2016 reflect the main-streamed political consideration of geoengineering in the United States, with Congressional representatives from both major political parties demonstrat-ing increasing receptivity over time to the idea of geoengineering, although for different reasons. As illustrated in the analysis of the previous chapters, it is clear that among scientists, even strong proponents of geoengineering solu-tions usually see them as a supplement, stopgap, or Plan B to emissions abate-ment. Nevertheless, as geoengineering increasingly mainstreams, there exists a risk that Plan B encroaches upon Plan A, especially as politicians embrace the technical solutions divorced from climate change as their *raison d'être*.

Geoengineering represents a radical departure from previous risks and the potential for a new existential crisis, novel in its global scale and the inten-tionality of the process. Despite its novelty and globality, the paradigm of governance is based upon existing structures and assumptions. It is assumed policy would facilitate and guide behavior of market economies in respon-sibly pursuing CDR as it becomes cost effective to do so in addition to or in place of traditional mitigation. The assumption of market-based policies and decisions weighed by costs places geoengineering largely under the control of the market. Yet, deployment at scale of CDR and/or albedo modification would require international cooperation and planning at a level at least equal to that which the international community has failed so far to reach in regard to meaningful mitigation through emission reduction measures. Moreover, the potential implementation of geoengineering does not supersede the need for international cooperation on emissions reductions; rather it adds a new level of necessary cooperation, rife with similar political challenges. The discourses that guide geoengineering's development are characterized by knowledge-seeking and faith in technological advancement within the scien-tific professional sphere and the politics of unsustainability within the politi-cal sphere. The counter-pressure is provided by the discourses and actions associated with environmentalism and radical life politics.

This research informs and engages with three levels of consideration regarding geoengineering's trajectory and its implications. First, in terms of social theory, the current resurgence of attention placed on geoengineering represents a retreat toward the mentality of simple modernity, as contrasted to the reflexivity emblematic of high modernity as seen in the environmen-tal movement (e.g., Giddens 1990; 1994). It represents a new layer to the "politics of unsustainability" (i.e., Blühdorn 2000; 2007; 2011; Blühdorn and Welsh 2007). However, while Blühdorn argues that we have entered an era of "post-ecologism," this research finds the move from ecologism to be more complex and dynamic with a persisting tension between simultaneous forces of ecological reflexivity and instrumentality. These tensions can not only be seen in the big-picture debates around climate change and geoengineering but also embodied in the ambiguity of policy reports and individual actors who, for example, advocate for geoengineering research, while remaining reluctant about its use and reiterating the primacy of mitigation. At the grass-roots level, environmental social movements, which have been engaged and active over the same time-period that geoengineering has been mainstreaming, remain the vanguard of reflexivity in the realm of climate politics, as will be discussed further.

Second, organizational dynamics as well as political-economic structures and conditions affect the trajectory of geoengineering. There are clear economic incentives among certain interests to promote geoengineering (Gunderson, Petersen and Stuart 2018; cf. Long and Scott 2013). There are also push factors embedded within institutional and professional cultures (cf. Long and Scott 2013). As the "inhabited institutions" concept within organizational sociology has elucidated, dynamics within organizations are complex and agentic individuals within them "possess varied, and sometimes crosscutting logics of action [. . .] on a continuum of almost purely universalistic to almost purely institutional" and "these actors are idiosyncratically endowed with interests" (Binder 2007, 567–568). Applied to geoengineering, particular components of professional culture among scientists and their commitment to organizational success influence the scientific community's treatment of geoengineering.

Most scientists engaged with geoengineering premise their position on underlying concerns regarding the seriousness of the threat of climate change and the corollary objective to maximize potential climate response options. The professional culture of research scientists also tends to promote the furthering of scientific knowledge as intrinsically valuable (cf. Stilgoe 2015, 16). The humanistic concern regarding climate change as well as the professional culture valuing scientific progress would constitute the more universal side of the continuum while the other end includes interests regarding agency, departmental, and personal research funding and prestige. A diverse array of individuals from a diverse array of institutions are active in shaping the trajectory of geoengineering. Actors include engineers, physicists, climatologists and ecologists as well as investors, philanthropists, social scientists, bureaucrats, and politicians from organizations that include for-profit corporations, foundations, academic research institutions, and government agencies. Crosscutting interests, tendencies and concerns, interact in affecting individual and organizational positions that influence the future of geoengineering.

Third, as examined empirically in the previous chapters, discursive trends, both strategic and incidental, significantly influence the reception and consideration of geoengineering as a potential response to climate change. In particular, in the genres of science policy reports and general audience journalism,

certain discursive trends contribute to the construction of legitimacy in relation to the field of geoengineering notwithstanding an explicit emphasis on
its novelty and risks. Within the political sphere, different ideologically-based
framings compete for legitimacy. The staging and framing of an emerging
technology like geoengineering affects its reception and political support,
while also reflecting values, interests, and concerns of the speakers.

Interactions between these social, cultural, and institutional influences as
well as the related forms of discursive presentation are multi-directional. The
politics of unsustainability and the corresponding move away from ecologism
influence the incentive structures at the political-economic level, which in
turn influences the discourse. From the other direction, the discursive trends
support the institutional structures that are advancing geoengineering and
reinforce or enact the politics of unsustainability.

Some individuals straddle the lines of interests, roles, and underlying
ideology. For example, David Keith is emblematic of being involved with
nearly every aspect of geoengineering in various roles and within organizations of differing types and purposes, while prolifically contributing to the
discourse of geoengineering. Keith is the most cited proponent of geoengineering within popular literature (see chapter 4), has served as a member of
the Royal Society working group that penned the influential 2009 report and
also served as a reviewer of the National Research Council's 2015 report (see
chapter 2), and has testified before Congress as an expert geoengineering
witness (see chapter 6). In terms of professional roles, Keith is a professor
that runs a research group on solar geoengineering at Harvard, but he is also
an entrepreneur, having founded Carbon Engineering, a for-profit corporation pursuing industrial-level CDR projects. Keith defines himself both as an
"environmentalist," stating "Wilderness has shaped my life," as well as "a
tinkerer and technophile" (Keith 2013, xiii-xiv). Keith embodies the ambiguity and complexity of advocating for the hubristic techno-fix of instrumental
control over nature, while practicing a personal version of ecological reflexivity, which he articulates in the preface of his manifesto on geoengineering
(Keith 2013, ix-xix).

Other influential actors involved in geoengineering also exemplify the
complexities, tensions, and ambiguities involved in geoengineering. For
instance, in contrast to Keith's enthusiastic advocacy, climate scientist and
geoengineering researcher Alan Robock has been particularly outspoken in
his concerns and reluctance about geoengineering (as discussed in chapters
4 and 6). He authored the article "20 Reasons Why Geoengineering May
Be a Bad Idea" (Robock 2008). He acted as a whistleblower on the CIA's
interest in geoengineering (Sample 2015). He unequivocally told members
of Congress that mitigation was paramount to addressing climate change and
that geoengineering absolutely cannot supplant it:

global warming is a serious problem and [. . .] mitigation, reduction of emissions, should be our primary response. We also need to do adaptation and learn to live with some of the climate change which is going to happen no matter what. [. . .] We can't hold geoengineering as a solution and allow that to reduce our push toward mitigation. It is never going to be a complete solution. We may need it in the event of an emergency, but let us not stop mitigation and wait and see if geoengineering would work. That is not the right strategy. (Congressional Hearing 2009a, 43, 87)

With his strong reservations on geoengineering, Robock's reluctant support for geoengineering research speaks to the nuance and ambiguities ingrained in geoengineering discourse, especially situated within the context of climate change politics and the unfolding reality of climate change itself. As mentioned in chapter 4, science journalist Sharon Begley wrote in *Newsweek* magazine: "In a sign of how dangerous global warming is starting to look and of how pitiful the world's efforts to control greenhouse gases are, even Robock—list [of 20 reasons geoengineering may be a bad idea] and all—hedges his bets" with geoengineering (Begley 2007).

Climate change is characterized by nonlinear unfolding of new risks, complicated by tipping points and feedback effects. The idea of a tipping point is that there is some threshold at which the climate system transitions significantly, and perhaps irreversibly, out of the existing equilibrium and toward a new one (see Lenton 2011). There is debate as to exactly what these tipping points may include and whether some may have been reached (e.g., see Revkin 2009; Levitan 2013). Feedback effects include environmental changes that result from climate change that, in turn, exacerbate climate change. For example, when ice sheets in the arctic melt due to the warming effects of climate change, and then that melting changes the composition more in favor of water (which absorbs heat) than ice (which reflects heat), causing the warming in the arctic region to proceed significantly faster than previous rates. These factors related to climate change are fitting analogies for geoengineering, which is proposed as an approach to address that climate change. Within the social world, there are tipping points and feedback cycles as well.

In regard to the establishment of legitimacy around the pursuit of geoengineering, one key tipping point with resultant feedback effects can be characterized around the upswell of geoengineering attention following Paul Crutzen's article (2006) in the journal *Climatic Change*, which advocated for consideration of albedo enhancement with stratospheric aerosols as a possible strategy to mitigate effects of climate change. This seminal article demonstrates the interconnections and interactions between scientific, political and popular discourse on geoengineering, as well as the profound influence one contribution can make. As discussed in chapter 4, it was following this article's publication that news media began covering geoengineering to an appreciable degree. However, not only did Crutzen's article indicate a move toward legitimacy for geoengineering spectators, but it inspired climate scientists to take geoengineering seriously. For example, Dr. Phillip Rasch, a climate scientist within the U.S. National Laboratories who has become an influential geoengineering researcher, states during the November 2017 congressional hearing on geoengineering that the Crutzen article "is what brought me into the field" (Congressional Hearing 2017b). Moreover, the NRC cites to Crutzen's article in explaining motivation for its examination of climate intervention in the seminal NRC *Reflecting Sunlight* report (National Research Council 2015b, 31). Finally, the 2021 NASEM *Reflecting Sunlight* "report is dedicated to Paul J. Crutzen (1933–2021) and Steve Rayner (1953–2020)" for being "pioneering researchers" who "made foundational contributions to solar geoengineering scholarship" (NASEM 2021, xiii).

As also discussed in chapter 4, there was contention regarding the publication of Crutzen's article promoting consideration of solar geoengineering, with the concern that it would be "irresponsible" to promote geoengineering as a possible response to climate change (Mark Lawrence, cited in Broad 2006). The compromise was reached in which Crutzen's editorial was published alongside counterargument editorials. As it turned out, however, Crutzen's article had a disproportionate effect compared to those counterarguments, with geoengineering gaining attention, prominence, and legitimacy as a result of it. Tracing the discourse of geoengineering from within scientific, popular and political discourse, a snowballing effect of legitimacy becomes apparent. Crutzen's article inspired new scientists to turn their attention to geoengineering. These scientists contributed to the National Academies' influential NRC report, among others. News media point to both the Crutzen article itself and the NRC report as indicating the mainstream legitimacy of geoengineering. Congress calls upon these scientists to provide testimony at hearings aimed at informing legislative policy on geoengineering. And so it goes.

CHANGING THE POLITICAL DEBATE

The notion that geoengineering could serve as a "hedge" or "Plan B" for confronting climate change realigns the parameters of political debate with the idea that there are a variety of potential responses to climate change. Increasing scientific attention to geoengineering has potential to fragment scientific resources, dilute the sense of consensus regarding the need for emission reductions, and open up new scientific and technical uncertainties which can be exploited by political interests hostile to emission reductions. Clive

Hamilton argues that in the 1990s, geoengineering proposals were regarded as a "distraction from the real task of reducing emissions" and that "almost all climate scientists took the view that the availability of an alternative to cutting emissions, even if manifestly inferior, would prove so alluring to political leaders that it would further undermine" emission reductions (2013a, 14–15). As seen in the previous chapter, this certainly was the case in regard to Republican Congressional Representatives on the House Committee on Science, Space, and Technology finding solar geoengineering to be particularly alluring.

The Solar Radiation Management Governance Initiative report notes the potential for climate engineering research to present "moral hazard" in which the perception of protection "against the potential consequences of climate change" may make people or governments "less likely to take the actions necessary to reduce greenhouse gas emissions" (2011, 20). The NRC Committee recognizes this risk of moral hazard, but contends that we have reached the point where "the severity of the potential risks from climate change appears to outweigh the potential risks from the moral hazard" in regard to geoengineering (National Research Council 2015b, 8).[1] Among the general public, Phil Macnaghten and Bronislaw Szerszynski found:

> Even though solar radiation management may be presented in good faith as "Plan B" ("Plan A" being continued effort at climate mitigation), there was a shared concern across the [focus] groups that its very availability as a technological option would weaken the commitment to climate mitigation—the well-known "moral hazard" argument, that today's "Plan B" would become tomorrow's "Plan A." (Macnaghten and Szerszynski 2013, 470)

Hence, mere consideration of geoengineering is perceived by some "as indirect permission to abandon serious efforts to cut emissions" (Specter 2012, 100). However, even among scientists working directly on geoengineering schemes, there is a strong reluctance toward the prospect of actually implementing geoengineering. Hugh Hunt, leader of the Cambridge team of the SPICE project, is quoted as saying "the last thing I would ever want is for the project I have been working on to be implemented. [. . .] If we have to use these tools, it means something on this planet has gone seriously wrong" (quoted in Specter 2012, 98). This statement is emblematic of the unique position of reluctance and ambiguity experienced by scientists working on geoengineering technologies. Naturally, research scientists are excited about advancing knowledge in a new field, but, as Hunt indicated, that does not translate into a desire to implement the technologies, which inherently carry risks and uncertainties.

As with many technical reports on climate engineering, the influential Royal Society (2009) and NRC (2015a; 2015b) reports are laden with caveats and identified uncertainties within their analyses of possible geoengineering options. The Royal Society states that "geoengineering of the Earth's climate is very likely to be technically possible. However, the technology to do so is barely formed, and there are major uncertainties regarding its effectiveness, costs, and environmental impacts" (Royal Society 2009, ix). The NRC report bluntly indicates: "There is significant potential for unanticipated, unmanageable, and regrettable consequences in multiple human dimensions from albedo modification at climate altering scales, including political, social, legal, economic, and ethical dimensions" (National Research Council 2015b, 148). While arguing that geoengineering is a necessary "Plan B," Lord Rees of the Royal Society, admits "Geoengineering would be an utter political nightmare" and that there "could be unintended side-effects" (as quoted in Jha 2013). The notion of a political nightmare is telling since one of the premises for possible geoengineering is the preexisting political failures to implement meaningful international carbon reductions and standards. Yet, a techno-fix cannot overcome the political challenges inherent in addressing a global environmental catastrophe, but rather adds a new layer.

The idea of a "Plan B" by definition implies that it is a back-up option, rather than a primary strategy. However, despite overwhelming scientific evidence and broad international acknowledgment that there is first and foremost a need to cut GHG emissions, the argument for a Plan B becomes stronger with each passing year as rising carbon emissions continue mostly unabated (e.g., Rees 2009, v; National Research Council 2015b, ix). In his overview of the American political response to climate change, Paul Harris explains how the response of several presidential administrations "to global warming was to call for more scientific research, which some interpreted as simply a recipe for pushing any requirement for US action well into the future" (2013, 67).[2] Calls to research geoengineering as a Plan B perpetuate this precedent of pushing meaningful climate mitigation policy further into the future as hope lingers for a technological solution. In the case of geoengineering research, pushing out the time horizon increases the probability of this Plan B eventually being implemented. The NRC report argues that, while they advise against deployment of albedo modification "at this time," research toward understanding and developing the technology should be pursued in case the point comes that such technology would be "useful" (National Research Council 2015b, e.g., 7–8, 49–54, 121). This leaves open the idea of an undefined future moment in which there would be some clarity of action to trigger Plan B, but with the details left for development.

Controlling discourses of the "future" has been noted as an important strategy in presenting and enhancing legitimacy in regard to contested or emerging technologies (Welsh 2000, 6; Selin 2007; Brown and Michael 2003). Ian Welsh notes that big science projects [. . .] typically have very long lead times which almost inevitably involve considerable areas of uncertainty. [. . .] The future invoked within such discourse typically emphasises positive collective outcomes for 'mankind' in the face of current uncertainties and doubts. (Welsh 2000, 6)

Relatedly, Nik Brown and Mike Michael identify the framing of "temporal representations of change and the future" and also note the "metaphor of the 'breakthrough'" as a "pervasive discursive method for organizing narratives about science" (Brown and Michael 2003, 7–8). This sort of positive framing of the technology's development can be seen in geoengineering discourse (as mentioned in chapter 2). In regard to geoengineering, however, the future is additionally invoked in another sense, as characterized by the "Plan B" and "climate emergency" framings, which project a future in need of saving.

Similar to the way geoengineering has moved from "taboo" to "serious" consideration in the scientific sphere, the same is seen in politics. As discussed in the previous chapter, the House of Representatives Science Committee over eight years of hearings has seen increasing support for geoengineering among members from both political parties. Subsequently, Andrew Yang, Democratic presidential candidate during the 2019–2020 primaries, further elevated the once taboo subject in public political discourse when he became the first national candidate to publicly take a stand in favor of geoengineering during his campaign. Yang included a policy statement on geoengineering in his campaign materials and discussed it during the CNN Town Hall on Climate Change on September 4, 2019 (Collinson 2019; Yang Campaign 2019). The movement of geoengineering from the margins toward mainstream deliberations on climate-change policy reflects and adds new dimensions to the preexisting theme of technological optimism in confronting climate change.

Discourse around how to address climate change has consistently included a technocratic and techno-enthusiast component. This is true even when geoengineering is bracketed from the equation. Optimism for the potential of decarbonization through innovation, a prime instantiation of ecological modernization theory's notion of dematerialization, is a central theme in climate discourse. Bill Gates' (2021) book on climate change, *How to Avoid a Climate Disaster,* exemplifies this optimism that innovation is the key to getting to zero emissions. Despite this central theme of optimism in regard to technological innovation, the title itself clearly indicates the consequences should the challenge not be met at high speed with full force, namely climate disaster. This risk leaves open the possible need of another type of technological innovation in regard to climate change. Climate disaster, of course, is a key theme in the narrative of geoengineering as a Plan B.

Despite the monumental risks involved, geoengineering resonates with a culture that is exuberant about technology solving the world's problems, even and especially those caused by technology in the first place (Blühdorn and Welsh 2007; Huesemann and Huesemann 2011). The idea of a technological fix through "climate engineering is intuitively appealing to a powerful strand of Western technological thinking and conservative politicking that sees no ethical or other obstacle to total domination of the planet" (Hamilton 2013a, 18). Geoengineering as a techno-fix to climate change is the most obvious manifestation of a social and political retreat from reflexivity back toward the technological faith and economic dogmas of simple modernity, characterized by "an instrumental approach to nature, faith in technological progress and abstract systems of expertise, and the exclusion of ambivalence and uncertainty" (Thorpe and Jacobson 2013, 100; cf. Giddens 1994, 5–7, 80–87).

The prospect of geoengineering as a potential technological solution offers a new strategy to repress the ontological insecurity that accompanies concerns of unmitigated climate change. Clive Hamilton contends: "Everyone is looking for an easy way out" and the "technofix of geoengineering [offers] a third way out" after the coping strategies of denialism and optimism (2013a, 107). However, geoengineering itself poses new global risks that present similar issues of ontological insecurity and existential crisis, literally a crisis of existence, to replace—or join—that related to the dangers of climate change itself. This sentiment can be seen in Phil Macnaghten and Bronislaw Szerszynski's focus group data in which subjects expressed fears of being part of a real-world global experiment with the potential to "destroy the Earth" (2013, 470).

Ian Welsh (2007) argues that technologies are promoted by scientific social movements and those that come to prevail have compatibility with social zeitgeist. However, publics are not homogenous, and reception of technologies is not straightforward. In his case study on atomic science, he finds that the public has always been ambivalent about nuclear technology, with varying levels of quiescence within the broader public, but always a presence of opposition and resistance (Welsh 2000). The same appears to be the case with geoengineering technology. David Keith, geoengineering advocate *par excellence*, commented in his testimony to Congress: "It is a healthy sign that a common first response to geoengineering is revulsion. It suggests we have learned something from past instances of techno-optimism and subsequent failures" (Congressional Hearing 2009a). Geoengineering enthusiast that he is, he then goes on to argue that "we must not over-interpret past experience" and that climate policy must include "sharp emissions cuts and clear-eyed research on SRM linked with the development of shared tools for managing it" (Keith statement at Congressional Hearing Congressional Hearing 2009a). While this statement pivots to conclude in support of advancing solar geoengineering technology, the starting point is notable with Keith pointing to the tendency of a recoiling response to the prospect of geoengineering, which makes room for the potential of countervailing forces to resist techno-cratic determinism.

GEOENGINEERING AND THE ECONOMY

Within the climate science community and policy arena, proposals for advancing climate engineering research and experiments are generally framed very explicitly as *not* an alternative to emission reductions. For example, the Royal Society report clearly states that

> the safest and most predictable method of moderating climate change is to take early and effective action to reduce emissions of greenhouse gases. No geoengineering method can provide an easy or readily acceptable alternative solution to the problem of climate change. (Royal Society 2009, ix)

The National Research Council explains: "There is no substitute for dramatic reductions in the emissions of CO_2 and other greenhouse gases to mitigate the negative consequences of climate change, and concurrently to reduce ocean acidification" (National Research Council 2015b, 2). In his forward to the Royal Society report, Lord Rees frames this premise by stating:

> nothing should divert us from the main priority of reducing global greenhouse gas emissions. But if such reductions achieve too little, too late, there will surely be pressure to consider a "plan B"—to seek ways to counteract the climatic effects of greenhouse gas emissions by "geoengineering." (Rees 2009, v)

According to the Royal Society's conclusions:

> The global failure to make sufficient progress on mitigation of climate change is largely due to social and political inertia, and this must be overcome if danger-ous climate change is to be avoided. If this proves not to be possible, geoengineering methods may provide a useful complement to mitigation and adaptation if they can be shown to be safe and cost effective. (Royal Society 2009, 57)

The NRC supports this idea that geoengineering would complement, not supersede, mitigation. As discussed in chapter 2, they refer to a "portfolio of climate responses" which could include mitigation, CDR and, perhaps, albedo modification (National Research Council 2015b, ix, 146). In the case of the latter, they emphasize:

> The less CO_2 that humans release to the atmosphere, the lower the environ-
> mental risk from the associated climate change and the lower the risk from any
> albedo modification that might be deployed as part of the strategy for address-
> ing climate change. It is widely recognized that the possibility of intervening
> in climate by albedo modification does not reduce the importance of efforts to
> reduce CO_2 emissions. (National Research Council 2015b, 36)

In any of the scenarios, then, whether or not geoengineering is integrated within the "portfolio of responses," mitigation remains essential.

In contrast to this careful caveated framing by scientific experts, for some financially interested parties, promoting geoengineering is a back-up to ped-dling skepticism as opposed to a good faith back-up for failures of timely emissions reductions. For instance, the oil and gas industry has been a major contributor to promoting climate change "skepticism" and, more recently, promoting geoengineering. Claire Parkinson explains "climate skeptics and friends of the fossil fuel industry have [. . .] discovered geoengineering" (2010, 15). For instance, ExxonMobil and some of its executives have been instrumental in the American Enterprise Institute, which has long worked "to deny the scientific consensus on climate change" and which now runs "one of the few funded policy centers on geoengineering" (Parkinson 2010, 15). As mentioned in chapter 6, one witness at the first congressional hear-ing on geoengineering was the codirector of AEI's Geoengineering Project, who was unique in his enthusiastic promotion of solar geoengineering and notably minimal discussion of mitigation. Exxon has also directly recruited engineers who, "ensconced in Exxon" have become influential on geoengi-neering, even influencing "'independent' reports into geoengineering, such as the 2007 NASA report on solar radiation management" (Hamilton 2013a, 78). Geoengineering in this instance constitutes a different kind of Plan B. Rather than the Plan B of trying to mitigate the potential extremes of climate change, for those deeply invested in the carbon-intensive energy economy, geoengineering is a Plan B—a back-up to the first strategic plan of climate change denial—to resist or defer the advancement of emission reductions.

It is not just oil and gas interests who are opposed to cutting emissions. Carbon emissions are deeply embedded into the worldwide industrial system of production. Constant growth is central to the global economic system and is necessary to avert economic crises within the system of competitive capitalism (Harvey 2010; Harman 2010). As discussed in chapter 1, the modern economic system has been characterized by environmental sociolo-gists as "the treadmill of production," in which production drives the growth of the global economy as well as the corresponding environmental effects (Gould, Pellow and Schnaiberg 2008). The treadmill of production theory emphasizes technology and production as the critical variables in explaining environmental impact starting in the third quarter of the twentieth century (Gould, Pellow and Schnaiberg 2008, 19). They propose that "contrary to classical and neoclassical economic theories that posit that consumer preferences determine the contour of markets, consumer behavior is consciously being shaped by industry" (Gould, Pellow and Schnaiberg 2008, 21).

Climate and geoengineering science policy reports tend to obliquely allude to this challenge, of industry and vested interests' influence over the market and climate politics. For example, the NRC report notes that "mitigation, although technologically feasible, has been difficult to achieve for political, economic, and social reasons that may persist well into the future" (2015b, 146). Similarly, the IPCC states that while there exist "multiple mitigation pathways that are likely to limit warming to below 2°C relative to preindustrial levels. [. . .] Implementing [the necessary GHG emission] reductions poses substantial technological, economic, social, and institutional challenges" (IPCC 2014, 90).

There have been significant advances in alternative energies in recent years. However, overcoming technical challenges to mitigation does not have a corollary effect in regard to economic, social, and institutional challenges. It rather appears that such technical advancements in energy alternatives may create further incentives for entrenched interests to resist the threat of transformation of the energy system. For instance, climate advocate, former Vice President Al Gore, argues that in response to both solar and wind power becoming increasingly affordable, "utilities are fighting back [. . .] by using their wealth and the entrenched political power they have built up over the past century" (2014, 87). Finding a techno-fix that might avert some or all pressures for carbon emission reductions would thus be a utopian reconciliation for the present capitalist fossil fuel-dominated economy and the environmental crisis we face from global climate change.

Not only does geoengineering promote the general political agenda of economic growth and the interests of the entrenched energy economy (cf. Gunderson, Petersen and Stuart 2018), but furthermore, various reports have linked advocates of geoengineering to specific financial interests. For example, Hamilton notes that a number of the expert scientists and policy influencers in regard to geoengineering are holders of related patents (Hamilton 2013a, 17, 75–84, 173; cf. Lukacs, Goldenberg and Vaughan 2013). Investors in private geoengineering ventures include oil tycoons such as "Murray Edwards, a Canadian oil billionaire with perhaps the largest financial stake in developing Alberta's tar sands" (Hamilton 2013a, 74; cf. Vidal 2018). Such industry investors might be expected to have a double-interest in the success of geoengineering as a techno-fix to climate change, especially if it can dampen calls for emission curtailment while making a profit in and of itself.

Looking at the example of Harvard University, which is a hub of influential geoengineering research, the Solar Geoengineering Research Program there has made the mindful decision to

> not accept donations from corporations, foundations, or individuals if the major-
> ity of their current profits or wealth come from the fossil fuel industry unless
> they can clearly demonstrate that they do not have a conflict of interest and
> present a strong track record of supporting efforts to address climate change.
> (Harvard's Solar Geoengineering Research Program 2021)

This policy represents a best practice of making efforts to avoid influence from fossil fuel interests in albedo modification research. While the specific geoengineering research program has this policy, however, fossil fuel interests do contribute funding to other programs and departments that may overlap with the researchers working on solar geoengineering. For example, an article published as part of the Harvard Project on Climate Agreements, written by David Keith along with co-authors Joshua Horton and Matthias Honegger, entitled "Implications of the Paris Agreement for Carbon Dioxide Removal and Solar Geoengineering," is notable in regard to the funders acknowledged. Among the funders listed for the project itself are the Enel Foundation and among funders listed for the "closely affiliated" Harvard Environmental Economics Program are BP, Chevron Services Company, Duke Energy Corporation, and Shell (Horton, Keith and Honegger 2016). Even academic research within Universities that are proactively making efforts at best practices to avoid influence of vested interests faces the challenge of encroachment of fossil fuel interests within the funding sphere.

Distinct from the oil and gas interests, one of the most important investors in geoengineering is not an oil but a silicone billionaire. Bill Gates is "the world's leading financial supporter of geoengineering research" (Hamilton 2013a, 74). Unlike the special interest investors mentioned above, Gates' interest in addressing climate change is clearly philanthropic and certainly not limited to geoengineering. Notably, Gates has been investing widely and systematically in climate change technology at numerous levels, including energy efficiency, carbon-sequestering concrete, meat alternatives, nuclear technology, as well as geoengineering. In terms of geoengineering, Gates has supported both CDR and solar geoengineering research through the Fund for Innovative Climate and Energy Research (FICER) which he has personally funded (Keith Group 2021). FICER funds have supported among others, the research of David Keith at Harvard University and Ken Caldeira at Stanford University (Keith Group 2021).

Other examples of wealthy individuals and groups supporting research include the Virgin Earth Challenge and the XPrize for Carbon Removal, both of which offer huge rewards for innovation in CDR. Interest in CDR is, of course, distinct from interest in solar geoengineering. As discussed in chapter 2, how people define the problem affects what solutions are fitting. For those who acknowledge anthropogenic climate change as emissions based, in addition to mitigation, CDR is a logical solution to pursue (whereas erstwhile climate-change deniers have been more enthusiastic about albedo modification). For instance, among private CDR supporters, Richard Branson exemplifies someone quite invested in carbon-producing enterprises, yet who acknowledges and shows concern for the causes and risks of climate change. In 2007 Branson collaborated on launching the Virgin Earth Challenge, offering a $25 million reward for new CDR proposals. While they have yet to pay out the reward because "no entries satisfied all the necessary prize criteria" (Virgin 2021), the challenge demonstrates the value seen in CDR and the belief that the market can induce needed solutions. The XPrize for Carbon Removal, funded by Elon Musk and the Musk Foundation goes further, offering $100 Million, "the largest incentive prize in history" (XPRIZE Foundation 2021).

The ability of powerful individuals and organizations to disproportionately influence the future of geoengineering is a recurring concern among proponents, critics, and commentators alike (e.g., National Research Council 2015b, ix, 9, 32, 33, 107, 122; Royal Society 2009, 39–40; Victor 2011, 196–197; IPCC 2014, 102; NASEM 2021, 148, 156). In contrast to the potential of life politics as a transformative and multilateral response to the climate crisis, geoengineering detracts from this democratic turn, regressing toward an expanded role of oligopoly in climate politics. The concept of geoengineering serves powerful political and economic interests by deflecting urgency in emissions reductions, facilitating the notion that growth may continue within the carbon-intensive energy economy, and by creating of a new market for high-tech, high-cost technological research and development.

Ryan Gunderson, Brian Peterson and Diana Stuart argue that geoengineering, particularly the most favored form of solar geoengineering, stratospheric aerosol injection, "supports economic priorities (and powerful financial actors), protects an inherently ecologically harmful social formation, and relegates the fundamental social-structural changes needed to actually address climate change" (2018, 14). Elsewhere they argue specifically that "the Plan B frame" demonstrates "the system-maintenance potential of geoengineering: addressing climate change without going about making the structural changes necessary to reduce emissions" (Gunderson, Stuart and Petersen 2018, 710). Since "emissions reductions, unlike geoengineering, are expensive, rely more on social-structural than technical changes, and are at odds with the current system (i.e., the current social system may be structurally incapable of significantly reducing emissions)," they "predict that geoengineering strategies, no matter how risky, will increasingly be considered principal means to combat climate change, perhaps even as alternatives to emissions reductions" (Gunderson, Petersen and Stuart 2018, 14). The mainstreaming trends identified in this analysis tend to reinforce the basis of this prediction. However, the potential of countervailing social forces to pose a challenge to this direction remains an open question. The opposite of geoengineering as an instrumental, technological response to climate change is not merely the absence of geoengineering. Rather it is the revitalization of life politics and environmental social movements.

THE COUNTER MOVE OF ENVIRONMENTALISM

Reflecting upon the emergence of the so-called self-regulating market following the Industrial Revolution and the social dislocation caused by it, Karl Polanyi wrote in 1944: "Human society would have been annihilated but for protective counter-moves which blunted the action of this self-destructive mechanism" (Polanyi [1944] 2001, 79.) According to Polanyi, social history of the 19th Century is that of a double movement: the development and expansion of markets but also the protective measures to check the market's impact on what he calls the "fictitious commodities" of land, labor, and money. It can be surmised that without these counter moves, the perils of the market would have been more devastating. Likewise, the modern environmental movement has served as a critical countervailing force in the era of advanced industrial society, endeavoring to protect social and environmental values put in jeopardy by economic values.

Modern environmentalist movements and environmental thought emerged in the mid-twentieth century, arguably coalescing in the space between the 1962 publication of Rachel Carson's groundbreaking book, *Silent Spring*, and the first Earth Day in 1970. This emergent environmentalism linked scientific knowledge with new social values and political mobilization. Scholarship on New Social Movements traced a shift in activism from traditional grievances, especially economic, to more cultural and social aims, which included the ecological movement (e.g., Habermas 1981). These movements, and forms of consciousness, posed a fundamental challenge to the modern view of economic growth and technological progress as goods in themselves.

The significance of environmentalism as a challenge to simple modernity was recognized sociologically by Anthony Giddens and centrally informed his conception of reflexive modernization (Beck, Giddens and Lash 1994; Giddens 2009; Giddens 1990; Giddens 1994). Other scholars of reflexive modernization, including world risk scholar Ulrich Beck (e.g., Beck 1997 [1993]; Beck 1992) as well as proponents of ecological modernization theory (e.g., Mol and Spaargaren 2000; Mol 2000), have also pointed to the significance of environmentalism in regard to reflexive modernization, as it encapsulates two key factors of reflexivity: impact science and social movements (McCright and Dunlap 2010, 103–104). For Giddens (e.g., 1990; 1994), environmentalism was a form of "life politics" that reopened the existential question of "how should we live?" This reflexive question, Giddens argued, had been closed down by modernity's compulsive drive toward economic growth and its one-dimensional focus on the technological control of nature (Giddens 1994, 10–11; Thorpe and Jacobson 2013, 104).

The rise of life politics, characterizing high modernity, meant that modernity's suppression of existential dilemmas would give way to new forms of reflexivity (Giddens 1990, 38–45; 1994, 7, 13, 42, 86, 90, 111; Thorpe and Jacobson 2013, 104–105, 108). However, Giddens also recognized that life politics came into contradiction with the dominant institutions of the capitalist state, which were not geared toward dealing with the kinds of reflexive questions regarding quality of life. Hence, he suggested that high modernity would require a new "double democratization," at the local and global levels (Giddens 1998, 70–78). The subsequent failure of modern society to achieve such institutional transformation, evident in the stagnation of efforts to negotiate international treaties or implement meaningful material practices that would sufficiently reduce carbon emissions, is suggestive of the concept of the "politics of unsustainability" and society's shift toward "post-ecologism" (Blühdorn and Welsh 2007; Blühdorn 2000; 2011).

Climate politics generally, and geoengineering specifically, have been illustrative of both the "politics of unsustainability" and the retreat from reflexivity. However, the broad trends of top-down influence guiding this trajectory of unsustainability are not immune to fissures from below, in the form of a return to life politics at the grassroots level. In recent years, grassroots activism representing radical ecologism, antithetical to post-ecologism, has involved a number of local and international campaigns, efforts, and actions. Specific examples include 350.org's international divestment campaign that has successfully encouraged individual and institutional investors to pull money out of fossil fuel industries, with "total value of institutions" that have divested surpassing $6 trillion according to the organization (Fossil Free 2018); "kayaktivists" in Seattle practicing civil disobedience to obstruct Shell's deep-sea oil rig on its way to Arctic drilling (Keim and Macalister 2015); and the concerted efforts of environmentalists in resisting the construction of new oil pipelines.

In terms of pipelines in the United States, working with other environmental activists, members of numerous Native American tribes have been leaders in standing up for environmental protection from oil pipelines. These have included the Keystone XL pipeline that became a focus of climate activism (Avery 2013),[3] the concerted efforts, led by Native Americans, in resisting the construction of the Dakota Access Pipeline at Standing Rock (Heim 2016), the Klamath tribe in Oregon in their own battle over the Jordan Cove LNG Pacific Connector Pipeline (Gentry and Marris 2018; Tribes 2017), and the "water protectors" of the Ojibwe nations in Northern Minnesota trying to protect their water, lands, and treaty rights from the Line 3 pipeline bringing oil from Alberta's tar sands (Regan 2021). In all of these cases, activists have shown a multi-pronged commitment to protecting local lands and waters from the pollution of leaks and bursts, the climate from additional investment in oil infrastructure and resulting emissions, and the cultural rights of local communities, which in these cases particularly include Native American tribes. Climate action movements represent a rekindling of a multi-faceted social reflexivity that poses a challenge to the hegemony of a "post-ecologist" paradigm.

While, for the most part, states have remained the key official actors in climate politics despite the limitations of state-based climate diplomacy (Harris 2013), broad grassroots environmental movements have served as the vanguard of life politics relevant to climate change and have been responsible for nudging governments forward on what progress has occurred on climate policy. The role of movements in climate politics, and their potential to affect norms and mores relevant to climate-change mitigation, is an important component of international climate politics. International climate negotiations have historically been challenged by the lack of appropriate institutional arrangements (Harris 2013; Victor 2011). However, where formal institutions are missing, informal institutions such as standards of behavior, norms and cultural values, are critically important (North 1990). To date, the existing achievements on climate mitigation efforts have largely been led by civil society movements (cf. Urry 2011, 111, 114). If we are to avoid the "Plan B" of geoengineering in response to climate change, environmental social movements will continue to be a big part of driving change in terms of policy and social values. Social movements help fill the vacuum created by the failure of formal international institutions to adequately regulate carbon consumption behavior.

Hence, social movements and life politics are essential to a transformation toward sustainability. This is not to discount the role of the structural, technological, and political changes that must be part of climate solutions, but rather highlight that these are interrelated and dependent on the life politics element to be successful. For instance, decarbonization technological advancements are certainly important. Without advancements in renewable energy like wind and solar as well as other lifestyle technologies like electric vehicles and meat alternatives, we cannot achieve sustainability. Technology alone, however, cannot achieve transformation. The public needs to take active roles in committing to sustainability and holding leadership accountable.

Likewise, enough pressure from the grassroots level can influence official policy, encouraging and guiding governmental leaders to be bolder in protecting the environment and promoting sustainability. For example, in the United States, the Biden administration made bold climate-change commitments as part of their campaign promises. Without pressure and demonstrable support in this direction from the public, however, achievement of climate goals will be limited. Moreover, activists at the informal, grassroots, and community levels can trailblaze solutions and transformations in their own terms as well.

Frank Fischer (2017) argues for local democratization, in the form of participatory governance, as an essential ingredient to climate mitigation. Relatedly, Andrew Stirling (2014) makes the case that to achieve sustainable solutions to the climate crisis, as well as other related environmental and social problems, the focus must be on *transformation* as opposed to merely *transitions*. Transition, the more typical form of political progress, results from efforts of incumbent political and economic elites guiding favorable policy "often driven by technological innovation, managed under orderly control, by incumbent structures according to tightly-disciplined frameworks for knowledge" (Stirling 2014, 1). In contrast, Stirling contends that transformation arises from the public and involves "social and technological innovations driven by diversely incommensurable knowledges, challenging incumbent structures" (2014, 1). This concept is relevant to the understanding of life politics as encompassing a plurality of values and practices that contribute toward cultural and material transformation. Stirling contends that geoengineering constitutes a "regressive" response to climate change "in the sense of being aligned with entrenched existing concentrations of power extending out from the energy sector" as opposed to a "progressive" transformation toward sustainable practices (Stirling 2014, 15).

Life politics, social movements, and participatory governance drive the potential for such transformation. This is the reverse of the technocratic approach of geoengineering. More than any other manifestation of life politics, environmentalism is necessarily *life* politics in a literal and biological sense. Life politics cannot just be concerned with the cultural question of how should we live, but it must also encompass the material question of how can we live sustainably.[4] In respect to environmentalism, then, life politics must change our relationship to the material world. Since modern society's primary relationship to the material world is intricately bound by systems of production and consumption, environmental life politics require reevaluation and transformation of these systems in a direction more compatible with the ecological systems with which they must necessarily interface and impact. Elsewise, in the context of climate change, the framing of geoengineering as a "Plan B" in case mitigation is "insufficient" becomes a discursive holding pattern until such time that the self-fulfilling prophecy triggers an enactment of this crisis script, which has become so embedded in the narratives explaining the legitimate role of geoengineering in society.

CONCLUDING SUMMATIONS ON
GEOENGINEERING DISCOURSE

The previous chapters have examined the discourse of geoengineering, with a focus on science policy reports, news media, and congressional hearings. These are three important genres with the ability to influence the trajectory of geoengineering's social and political acceptance as well as the substantive practices related to it, including the course of research and potential development of proposed technologies. Science policy reports on geoengineering from elite scientific societies have both reflected and promoted the mainstreaming of geoengineering (even while the authors maintain reluctancy toward deployment of solar geoengineering). Discursive strategies within such reports construct legitimacy and contribute to the mainstreaming of geoengineering within scientific, political, and public discourse. News coverage of geoengineering has increased since 2006, coinciding with important publications from the scientific community, with scientific publications used to indicate the mainstreaming of geoengineering as well as offering topical insight. Recurrent narratives within popular media also contribute to the mainstreaming of geoengineering through presenting its trajectory as moving from fringe origins to serious consideration. Over the course of four congressional hearings on the subject, geoengineering has increasingly garnered political support from both major political parties in the United States, but for different reasons and with different interpretations of the role it might play in climate policy. This study has also demonstrated the role of certain geoengineering researchers and advocates in influencing the deliberation and presentation of geoengineering within science publications, popular media, and policy discourse. These three genres reinforce one another in reflecting and advancing the mainstreaming of geoengineering.

Geoengineering discourse is comparable to that of other emerging technologies in that framings, narratives, and discursive strategies compete for influencing the terms of deliberation and future prospects of the technology's development. Geoengineering is distinct from other emerging technologies, however, in both its scope of possibilities and premise. In terms of possibilities, as previously mentioned, even proponents concede the global risks of dramatically changing or even extinguishing life on Earth. In terms of premise, geoengineering is envisioned as a possible response to climate change, itself imbued with global existential risks (Beck 2009; Thorpe and Jacobson 2013). While proponents of biotechnologies or nanotechnologies, for instance, identify potential for their research to address global problems, these problems are qualitatively and quantitatively distinct from climate change. The magnitude of climate change and its worst-case scenarios create a context in which otherwise outlandish or inconceivable geoengineering proposals have become mainstreamed, garnering serious consideration by scientists and governments. As illustrated in this analysis of geoengineering, consideration of the inconceivable is further aided by controlling the representation of this technology through discourse.

NOTES

1. The Royal Society, in contrast, indicates a level of uncertainty in regard to the moral hazard concern, speculating that rather than simply decreasing motivation for mitigation, "it is possible that geoengineering actions could galvanise people into demanding more effective mitigation action" and therefore calls for additional research into the risk of moral hazard as related to geoengineering (2009: 39).

2. In context, Harris was specifically referring to the Reagan and George H. W. Bush administrations.

3. After years of dispute and politicization, the Biden administration again revoked the permits for the Keystone XL pipeline with the "Executive Order on Protecting Public Health and the Environment and Restoring Science to Tackle the Climate Crisis" signed on January 20, 2021.White House. 2021a. *Executive Order on Protecting Public Health and the Environment and Restoring Science to Tackle the Climate Crisis*, January 20 Retrieved February 23, 2020 (https://www.whitehouse .gov/briefing-room/presidential-actions/2021/01/20/executive-order-protecting-public-health-and-environment-and-restoring-science-to-tackle-climate-crisis/).

4. This conception of *life* politics originated in conversation with Charles Thorpe.

References

Achenbach, Joel. 2015. "Short-Term Fixes for Longterm Climate Problems? Not So Fast, Experts Say." *The Washington Post*, February 9. https://www.washington-post.com/national/health-science/upcoming-report-on-technological-fixes-for-climate-change-adds-to-debate/2015/02/07/07bbe150-ac6e-11e4-9c91-e9d2f9fde644_story.html.

Adam, Barbara. 1998. *Timescapes of Modernity: The Environment and Invisible Hazards*. London: Routledge.

Asilomar Scientific Organizing Committee. 2010. *The Asilomar Conference Recommendations on Principles for Research into Climate Engineering Techniques*. Washington, DC: Climate Institute.

Avery, Samuel. 2013. *The Pipeline and the Paradigm: Keystone XL, Tar Sands, and the Battle to Defuse the Carbon Bomb*. Washington, DC: Ruka Press.

Baskin, Jeremy. 2019. *Geoengineering, the Anthropocene and the End of Nature*. Cham: Palgrave Macmillan.

Beaumont, Peter. 2018. "Scientists Suggest a Giant Sunshade in the Sky Could Solve Global Warming." *The Guardian*, April 5. https://www.theguardian.com/global-development/2018/apr/05/scientists-suggest-giant-sunshade-in-sky-could-solve-global-warming.

Beck, Ulrich. 1992. *Risk Society: Towards a New Modernity*. London: Sage Publications.

Beck, Ulrich. 1997. "Subpolitics." *Organization & Environment* 10 (1): 52–65.

Beck, Ulrich. 1997 [1993]. *The Reinvention of Politics*. Cambridge: Polity Press.

Beck, Ulrich. 2009. *World at Risk*. Translated by Ciaran Cronin. Malden: Polity.

Beck, Ulrich, Anthony Giddens and Scott Lash. 1994. *Reflexive Modernization: Politics, Tradition and Aesthetics in the Modern Social Order*. Stanford, CA: Stanford University Press.

Begley, Sharon. 1991. "On The Wings Of Icarus." *Newsweek*, May 19, 1991. https://www.newsweek.com/wings-icarus-203748.

Begley, Sharon. 2007. "'Geo-Engineering' Climate Change." *Newsweek*, November 22. https://www.newsweek.com/geo-engineering-climate-change-96713.

Begley, Sharon. 2011. "Begley: Problems with 'Geo-Engineering' Plans." *Newsweek*, January 30. https://www.newsweek.com/begley-problems-geo-engineering-plans -66709.

Bell, Daniel. 1973. *The Coming of Post-Industrial Society: A Venture in Social Forecasting*. New York: Basic Books.

Bellamy, Rob. 2013. "Framing Geoengineering Assessment." *Geoengineering Our Climate Working Paper and Opinion Article Series*. https://geoengineeringo urclimate.files.wordpress.com/2013/12/bellamy-2013-framing-geoengineering -assessment-click-for-download.pdf.

Bellamy, Rob, Jason Chilvers, Naomi E. Vaughan and Timothy M. Lenton. 2012. "A Review of Climate Geoengineering Appraisals." *Wiley Interdisciplinary Reviews: Climate Change* 3: 597–615.

Bellamy, Rob, Jason Chilvers, Naomi E. Vaughan and Timothy M. Lenton. 2013. "'Opening Up' Geoengineering Appraisal: Multi-Criteria Mapping of Options for Tackling Climate Change." *Global Environmental Change* 23: 926–937.

Biello, David. 2015. "Blocking the Sun Is No Plan B for Global Warming." *Scientific American*, December 9. https://www.scientificamerican.com/article/blocking-the -sun-is-no-plan-b-for-global-warming/.

Biermann, Frank and Klaus Dingworth. 2004. "Global Environmental Change and the Nation State." *Global Environmental Politics* 4 (1): 1–22.

Binder, Amy J. 2007. "For Love and Money: Organizations' Creative Responses to Multiple Environmental Logics." *Theory and Society* 36 (6): 547–571.

Black, Richard. 2012a. "Climate 'Tech Fixes' Urged for Arctic Methane." *BBC News*, March 17. www.bbc.co.uk/news/science-environment-17400804.

Black, Richard. 2012b. "Geoengineering: Risks and Benefits." *BBC News*, August 24. www.bbc.co.uk/news/science-environment-19371833.

Blühdorn, Ingolfur. 2000. *Post-Ecologist Politics: Social Theory and the Abdication of the Ecologist Paradigm*. London, New York: Routledge.

Blühdorn, Ingolfur. 2007. "Sustaining the Unsustainable: Symbolic Politics and the Politics of Simulation." *Environmental Politics* 16 (2): 251–275.

Blühdorn, Ingolfur. 2011. "The Politics of Unsustainability: COP15, Post-Ecologism, and the Ecological Paradox." *Organization & Environment* 24 (1): 34–53.

Blühdorn, Ingolfur and Ian Welsh. 2007. "Eco-Politics Beyond The Paradigm of Sustainability: A Conceptual Framework and Research Agenda." *Environmental Politics* 16 (2): 185–205.

Board of Editors. 2015. "A Hacker's Guide to Planet Cooling." *Scientific American*, January: 10.

Bracmort, Kelsi and Richard K. Lattanzio. 2013. *Geoengineering: Governance and Technology Policy*. Washington, DC: Congressional Research Service.

Broad, William J. 2006. "How to Cool a Planet (Maybe)." *The New York Times*, June 27. https://www.nytimes.com/2006/06/27/science/earth/27cool.html.

Brogan, Jacob. 2016. "Your Geoengineering Cheat Sheet." *Slate*, January 6. https://slate.com/technology/2016/01/geoengineering-101-a-cheat-sheet-to-the-terminology-the-key-players-and-more.html.

Brown, Nik, Alison Kraft and Paul Martin. 2006. "The Promissory Pasts of Blood Stem Cells." *BioSocieties* 1: 329–348.

Brown, Nik and Mike Michael. 2002. "From Authority to Authenticity: The Changing Governance of Biotechnology." *Health, Risk & Society* 4 (3): 259–272.

Brown, Nik and Mike Michael. 2003. "A Sociology of Expectations: *Retrospecting Prospects* and *Prospecting Retrospects.*" *Technology Analysis & Strategic Management* 15 (1): 3–18.

Buell, Frederick. 2003. *From Apocalypse to Way of Life: Environmental Crisis in the American Century*. New York: Routledge.

Canon, Gabrielle. 2021. "Wildfire Fighters Advance Against Biggest US Blaze Amid Dire Warnings." *The Guardian*, August 2. https://www.theguardian.com/us-news/2021/aug/02/us-wildfires-bootleg-fire-climate-change.

Carr, Ada. 2015. "Split Decisions on Geoengineering Emerge During Climate Talks." *weather.com*. Retrieved July 3, 2017. https://weather.com/news/climate/news/geoengineering-paris-climate-conference-talks.

Carson, Rachel. 2002. *Silent Spring*. Boston: Houghton Mifflin.

Carton, Wim. 2021. "Carbon Unicorns and Fossil Futures." In *Has It Come to This? The Promises and Perils of Geoengineering on the Brink*, edited by J. P. Saspinski, Holly Jean Buck and Andreas Malm, 34–49. New Brunswick: Rutgers University Press.

Ceballos, Gerardo, Paul R. Ehrlich and Rodolfo Dirzo. 2017. "Biological Annihilation via the Ongoing Sixth Mass Extinction Signaled by Vertebrate Population Losses and Declines." *Proceedings of the National Academy of Sciences* 114 (30): E6089.

Chang, Gordon and Hugh Mehan. 2008. "Why We Must Attack Iraq: Bush's Reasoning Practices and Argumentation System." *Discourse & Society* 19 (4): 453–482.

Choi, Charles. 2012. "Asteroid Dust Could Fight Climate Change on Earth." *LiveScience*, September 29. http://www.livescience.com/23553-asteroid-dust-geoenineering-global-warming.html.

Collinson, Stephen. 2019. "What Happened during CNN's Climate Town Hall and What it Means for 2020." *CNN*, September 5. https://www.cnn.com/2019/09/05/politics/climate-town-hall-highlights/index.html.

Committee on Science and Technology. 2010. *Geoengineering: Hearing Before the Committee on Science and Technology*. Washington, DC: House of Representatives.

Committee on Science, Engineering, and Public Policy (U.S.). 1992. *Policy Implications of Greenhouse Warming: Mitigation, Adaptation, and the Science Base*. Washington, DC: National Academy Press.

Commoner, Barry. 1971. *The Closing Circle: Nature, Man and Technology*. New York: Alfred A. Knopf.

Congressional Hearing. 2009a. "United States Congress: House Committee on Science and Technology." *Geoengineering: Assessing the Implications of Large-Scale Climate Intervention*. Hearing. 111th Congress. 1st Session, November 5.

Congressional Hearing. 2009b. "United States Congress: House Subcommittee on Energy and Environment." *Monitoring, Measurement, and Verification of Greenhouse Gas Emissions, Parts I and II.* Hearing. 111th Congress. 1st Session.

Congressional Hearing. 2010a. "United States Congress: House Subcommittee on Energy and Environment." *Geoengineering II: The Scientific Basis and Engineering Challenges.* Hearing. 111th Congress. 2nd Session, February 4.

Congressional Hearing. 2010b. "United States Congress: House Committee on Science and Technology." *Geoengineering III: Domestic and International Research Governance.* Hearing. 111th Congress. 2nd Session.

Congressional Hearing. 2013. "United States Congress: House Committee on Science, Space, and Technology: Subcommittee on Environment." *Policy Relevant Climate Issues in Context.* Hearing. 113th Congress.

Congressional Hearing. 2014. "United States Congress: House Subcommittee on Energy and Subcommittee on Environment." *Science of Capture and Storage: Understanding EPA's Carbon Rules.* Hearing. 113th Congress. 2nd Session, March 12.

Congressional Hearing. 2017a. "United States Congress: House Committee on Science, Space, and Technology." *Climate Science: Assumptions, Policy Implications, and the Scientific Method.* Hearing. 115th Congress, 1st Session, March 29.

Congressional Hearing. 2017b. "United States Congress: House Subcommittee on Environment and House Subcommittee on Energy." *Geoengineering: Innovation, Research, and Technology.* Hearing. 115th Congress. 1st Session, November 8.

Conover, John H. 1966. "Anomalous Cloud Lines." *Journal of the Atmospheric Sciences* 23: 778–785.

Corner, Adam, Karen Parkhill and Nick Pidgeon. 2011. "'Experiment Earth?' Reflections on a Public Dialogue on Geoengineering." *Working Paper.* Cardiff University.

Corner, Adam, Karen Parkhill, Nick Pidgeon and Naomi E. Vaughan. 2013. "Messing with Nature? Exploring Public Perceptions of Geoengineering in the UK." *Global Environmental Change* 23: 938–947.

Craik, Neil and William C. G. Burns. 2019. "Climate Engineering Under the Paris Agreement." *Environmental Law Institute* 43 (12): 11113–11129.

Cressey, Daniel. 2012. "Cancelled Project Spurs Debate over Geoengineering Patents." *Nature* 485: 429.

Crutzen, Paul J. 2006. "Albedo Enhancement By Stratospheric Sulfur Injections: A Contribution to Resolve a Policy Dilemma?" *Climatic Change* 77: 211–219.

Da-Allada, C. Y., E. Baloïtcha, E. A. Alamou, F. M. Awo, F. Bonou, Y. Pomalegni, E. I. Biao, E. Obada, J. E. Zandagba, S. Tilmes and P. J. Irvine. 2020. "Changes in West African Summer Monsoon Precipitation Under Stratospheric Aerosol Geoengineering." *Earth's Future* 8 (7): e2020EF001595.

Dasgupta, Partha. 2007. "A Challenge to Kyoto: Standard Cost–Benefit Analysis May Not Apply to the Economics of Climate Change." *Nature* 449: 143–144.

Davenport, Coral. 2015. "Nations Approve Landmark Climate Accord in Paris." *The New York Times*, December 12. https://www.nytimes.com/2015/12/13/world/europe/climate-change-accord-paris.html.

Dickens, Peter. 1999. "Life Politics, the Environment and the Limits of Sociology." In *Theorising Modernity: Reflexivity, Environment and Identity in Giddens' Social Theory*, edited by Martin O'Brien, Sue Penna and Colin Hay, 98–120. London: Longman.

Dryzek, John S. 2005. *The Politics of the Earth: Environmental Discourses*. Oxford: Oxford University Press.

Dunlap, Riley E., Aaron M. McCright and Jerrod H. Yarosh. 2016. "The Political Divide on Climate Change: Partisan Polarization Widens in the U.S." *Environment: Science and Policy for Sustainable Development* 58 (5): 4–23.

Ellison, Jesse. 2009. "Geo-Engineering: Quick, Cheap Way to Cool Planet?" *Newsweek*, November 27. https://www.newsweek.com/geo-engineering-quick -cheap-way-cool-planet-76697.

Environmental Pollution Panel. 1965. *Restoring the Quality of Our Environment*. Washington, DC: President's Science Advisory Committee.

ETC Group. 2010. *Geopiracy: The Case Against Geoengineering*, Ottawa. http://www .etcgroup.org/sites/www.etcgroup.org/files/publication/pdf_file/ETC_geopiracy _4web.pdf.

European Commission. 2017. *G20 Leaders' Declaration: Shaping an Interconnected World*. Hamburg. http://europa.eu/rapid/press-release_STATEMENT-17-1960_en .htm.

Fairclough, Norman. 2000. *New Labour, New Language?* New York: Routledge.

Fairclough, Norman and Isabela Fairclough. 2012. *Political Discourse Analysis*. New York: Routledge.

Farber, Dan. 2015. "Does the Paris Agreement Open the Door to Geoengineering?" *Berkelely Blogs*. http://blogs.berkeley.edu/2015/12/14/does-the-paris-agreement -open-the-door-to-geoengineering/.

Felt, Ulrike and Maximilian Fochler. 2010. "Machineries for Making Publics: Inscribing and De-scribing Publics in Public Engagement." *Minerva* 48: 219–238.

Fialka, John. 2020. "U.S. Geoengineering Research Gets a Lift with $4 Million from Congress." *Science*, January 23. https://www.science.org/content/article/us -geoengineering-research-gets-lift-4-million-congress.

Fischer, Douglas. 2009. "Engineering the Planet to Dodge Global Warming." *Scientific American*, November 12. https://www.scientificamerican.com/article/ copenhagen-consequences-geoengineering-climate-change/.

Fischer, Frank. 2017. *Climate Crisis and the Democratic Prospect: Participatory Governance in Sustainable Communities*. New York: Oxford University Press.

Fleming, James. 2010. *Fixing the Sky: The Checkered History of Weather and Climate Control*. New York: Columbia University Press.

Fossil Free. 2018. "2018 is Looking Like the Year of Mass Divestment." Accessed April 18, 2018. https://gofossilfree.org/2018-is-looking-like-the-year-of-mass -divestment/.

Foster, John Bellamy. 2002. *Ecology Against Capitalism*. New York: Monthly Review Press.

Foster, John Bellamy. 2011. "Capitalism and the Accumulation of Catastrophe." *Monthly Review* 63 (7): 1–17.

Foster, John Bellamy, Richard York and Brett Clark. 2009. "The Midas Effect: A Critique of Climate Change Economics." *Development and Change* 40 (6): 1085–1097.

Foucault, Michel. 2010. *The Archaeology of Knowledge and the Discourse on Language*. Translated by A. M. Sheridan Smith. New York: Vintage Books.

Fountain, Henry. 2014. "Climate Tools Seek to Bend Nature's Path." *The New York Times*, November 9. https://www.nytimes.com/2014/11/10/science/earth/climate-tools-seek-to-bend-natures-path.html.

Fountain, Henry. 2015. "Panel Urges Research on Geoengineering as a Tool Against Climate Change." *The New York Times*, February 10. https://www.nytimes.com/2015/02/11/science/panel-urges-more-research-on-geoengineering-as-a-tool-against-climate-change.html.

Fountain, Henry. 2021a. "5 Takeaways from the Major New U.N. Climate Report." *The New York Times*, August 9. https://www.nytimes.com/2021/08/09/climate/un-climate-report-takeaways.html.

Fountain, Henry. 2021b. "Hotter Summer Days Mean More Sierra Nevada Wildfires, Study Finds." *The New York Times*, November 17. https://www.nytimes.com/2021/11/17/climate/climate-change-wildfire-risk.html.

Fourcade, Marion. 2011. "Cents and Sensibility: Economic Valuation and the Nature of 'Nature'." *American Journal of Sociology* 116 (6): 1721–1777.

Friedman, Lisa. 2020. "Biden Introduces His Climate Team." *The New York Times*, December 19. https://www.nytimes.com/2020/12/19/climate/biden-climate-team.html.

Gardiner, Stephen M. 2011. "Some Early Ethics of Geoengineering the Climate: A Commentary on the Values of the Royal Society Report." *Environmental Values* 20 (2): 163–188.

Gentry, Don and Emma Marris. 2018. "The Next Standing Rock? A Pipeline Battle Looms in Oregon." *The New York Times*, March 8. https://www.nytimes.com/2018/03/08/opinion/standing-rock-pipeline-oregon.html.

Giddens, Anthony. 1981. *A Contemporary Critique of Historical Materialism, Vol. 1: Power, Property and the State*. Berkeley: University of California Press.

Giddens, Anthony. 1990. *The Consequences of Modernity*. Stanford: Stanford University Press.

Giddens, Anthony. 1991. *Modernity and Self-Identity*. Stanford: Stanford University Press.

Giddens, Anthony. 1994. *Beyond Left and Right*. Stanford: Stanford University Press.

Giddens, Anthony. 1998. *The Third Way: The Renewal of Social Democracy*. Cambridge: Polity Press.

Giddens, Anthony. 2009. *The Politics of Climate Change*. Cambridge: Polity Press.

Goodell, Jeff. 2010. *How to Cool the Planet: Geoengineering and the Audacious Quest to Fix Earth's Climate*. Boston: Houghton Mifflin Harcourt.

Gore, Al. 2014. "The Turning Point: New Hope for the Climate." *Rolling Stone*, July 3–17 (1212/1213): 84–93.

Gould, Kenneth A., David N. Pellow and Allan Schnaiberg. 2008. *The Treadmill of Production: Injustice and Unsustainability in the Global Economy.* Boulder: Paradigm Publishers.

Gould, Stephen Jay. 1981. "Evolution as Fact and Theory." *Discover* 2: 34–37.

Gunderson, Ryan, Brian Petersen and Diana Stuart. 2018. "A Critical Examination of Geoengineering: Economic and Technological Rationality in Social Context." *Sustainability* 10 (1): 269–290.

Gunderson, Ryan, Diana Stuart and Brian Petersen. 2018. "The Political Economy of Geoengineering as Plan B: Technological Rationality, Moral Hazard, and New Technology." *New Political Economy* 24 (5): 696–715.

Guterl, Fred. 2009. "Radical Ways to Cool the Planet." *Newsweek*, April 17. https://www.newsweek.com/radical-ways-cool-planet-77591.

Habermas, Jüergen. 1981. "New Social Movements." *Telos* Fall 1981 (49): 33–37.

Halstead, John. 2018. "Stratospheric Aerosol Injection Research and Existential Risk." *Futures* 102: 63–77.

Hamilton, Clive. 2013a. *Earthmasters: The Dawn of the Age of Climate Engineering.* New Haven: Yale University Press.

Hamilton, Clive. 2013b. "Geoengineering: Can We Save the Planet by Messing with Nature?" In *Democracy Now!*, edited by Amy Goodman, May 20. https://www.democracynow.org/2013/5/20/geoengineering_can_we_save_the_planet.

Hamilton, Clive. 2013c. "Geoengineering: Our Last Hope, or a False Promise?" *The New York Times*, May 27. https://www.nytimes.com/2013/05/27/opinion/geoengineering-our-last-hope-or-a-false-promise.html.

Hanley, Charles J. 2010. "Geoengineering Debate Surfaces As UN Climate Change Talks In Cancun Falter." *Huffington Post* December 4. http://www.huffingtonpost.com/2010/12/06/geoengineering-debate-sur_n_792409.html.

Hansen, James. 2009. *Storms of my Grandchildren: The Truth About the Coming Climate Catastrophe and Our Last Chance to Save Humanity.* New York: Bloomsbury.

Harman, Chris. 2010. *Zombie Capitalism: Global Crisis And The Relevance Of Marx.* Chicago: Haymarket Books.

Harris, Paul G. 2013. *What's Wrong with Climate Politics and How to Fix It.* Cambridge: Polity Press.

Harvard's Solar Geoengineering Research Program. 2021. "Funding." Accessed February 24, 2021. https://geoengineering.environment.harvard.edu/funding.

Harvey, David. 2010. *The Enigma of Capital: And the Crises of Capitalism.* Oxford: Oxford University Press.

Healey, Peter. 2014. "The Stabilisation of Geoengineering: Stabilising the Inherently Unstable?" *Climate Geoengineering Governance Working Paper Series* 015.

Heim, Joe. 2016. "Showdown over Oil Pipeline Becomes a National Movement for Native Americans." *The Washington Post*, September 7. https://www.washingtonpost.com/national/showdown-over-oil-pipeline-becomes-a-national-movement-for-native-americans/2016/09/06/ea0cb042-7167-11e6-8533-6b0b0ded0253_story.html.

Hilgartner, Stephen. 2000. *Science on Stage: Expert Advice as Public Drama*. Stanford: Stanford University Press.

Horton, Joshua B., David W. Keith and Matthias Honegger. 2016. "Implications of the Paris Agreement for Carbon Dioxide Removal and Solar Geoengineering." *Viewpoints*. Harvard Project on Climate Agreements.

House of Commons Science and Technology Committee. 2010. *The Regulation of Geoengineering*. HC 221, London.

House, Tamzy J., James B. Near, Jr., William B. Shields, Ronald J. Celentano, David M. Husband, Ann E. Mercer and James E. Pugh. 1996. *Weather as a Force Multiplier: Owning the Weather in 2025, Air Force 2025*. Accessed January 16, 2014. http://csat.au.af.mil/2025/volume3/vol3ch15.pdf.

Howell, Katie. 2010. "Climate: Scientists Call for Interagency Geoengineering Research Program." *Environment and Energy Daily*, February 5.

Huesemann, Michael and Joyce Huesemann. 2011. *Techno-Fix: Why Technology Won't Save Us or the Environment*. Gabriola Island: New Society Publishers.

Hulme, Mike. 2014. *Can Science Fix Climate Change?* Cambridge: Polity.

Huntingford, Chris and Lina M. Mercado. 2016. "High Chance that Current Atmospheric Greenhouse Concentrations Commit to Warmings Greater than 1.5 °C over Land." *Nature Scientific Reports* 6: 1–7.

Inslee, Jay. 2017. "United States Climate Alliance Adds 10 New Members to Coalition Committed to Upholding the Paris Accord." Accessed July 22, 2017. http://governor.wa.gov/news-media/united-states-climate-alliance-adds-10-new -members-coalition-committed-upholding-paris.

IPCC, Intergovernmental Panel on Climate Change. 1990. "Climate Change: The IPCC Response Strategies." *Working Group III Contribution to the First Assessment Report of the Intergovernmental Panel on Climate Change*.

IPCC, Intergovernmental Panel on Climate Change. 1995. "Climate Change 1995: Impacts, Adaptations and Mitigations of Climate Change: Scientific Analyses." *Contribution of Working Group II to the Second Assessment Report of the Intergovernmental Panel on Climate Change*. Cambridge: Cambridge University Press.

IPCC, Intergovernmental Panel on Climate Change. 2001. "Climate Change 2001: Mitigation." *Working Group III Contribtution to the Third Assessment Report of the Intergovernmental Panel on Climate Change*. Cambridge: Cambridge University Press.

IPCC, Intergovernmental Panel on Climate Change. 2007. "Climate Change 2007: Mitigation of Climate Change." *Contribution of Working Group III to the Fourth Assessment Report of the Intergovernmental Panel on Climate Change*. Cambridge: Cambridge University Press.

IPCC, Intergovernmental Panel on Climate Change. 2013. "Climate Change 2013: The Physical Science Basis." *Working Group I Contribution to the Fifth Assessment Report of the Intergovernmental Panel on Climate Change*. Cambridge: Cambridge University Press.

IPCC, Intergovernmental Panel on Climate Change. 2014a. "Climate Change 2014: Mitigation of Climate Change." *Working Group III Contribution to the*

Fifth Assessment Report of the Intergovernmental Panel on Climate Change. Cambridge: Cambridge University Press.

IPCC, Intergovernmental Panel on Climate Change. 2014b. "Climate Change 2014: Synthesis Report. Contribution of Working Groups I, II and III to the Fifth Assessment Report of the Intergovernmental Panel on Climate Change." Geneva.

IPCC, Intergovernmental Panel on Climate Change. 2018. "Summary for Policymakers." *Global Warming of 1.5°C. An IPCC Special Report on the Impacts of Global Warming of 1.5°C above Pre-industrial Levels and Related Global Greenhouse Gas Emission Pathways, in the Context of Strengthening the Global Response to the Threat of Climate Change, Sustainable Development, and Efforts to Eradicate Poverty.*

IPCC, Intergovernmental Panel on Climate Change. 2021. "Climate Change 2021: The Physical Science Basis. Contribution of Working Group I to the Sixth Assessment Report of the Intergovernmental Panel on Climate Change."

Jacobson, Brynna. 2018. "Constructing Legitimacy in Geoengineering Discourse: The Politics of Representation in Science Policy Literature." *Science as Culture* 27 (3): 322–348.

Jain, P. C. 1993. "Earth-Sun System Energetics and Global Warming." *Climatic Change* 24 (3): 271–272.

Jandt, Randi and Alison York. 2021. "Wildfire Is Transforming Alaska and Amplifying Climate Change." *Scientific American*, October 1. https://www .scientificamerican.com/article/wildfire-is-transforming-alaska-and-amplifying -climate-change/.

Janetsky, Megan and Matthew Kelly. 2018. "Trump Picks Top Koch Recipient for Secretary of State." *OpenSecrets News*: April 26. https://www.opensecrets.org/ news/2018/04/trump-picks-top-koch-recipient-for-secretary-of-state/.

Jasanoff, Sheila. 2015. "Future Imperfect: Science, Technology, and the Imaginations of Modernity." In *Dreamscapes of Modernity: Sociotechnical Imaginaries and the Fabrication of Power*, edited by Sheila Jasanoff and Sang-Hyun Kim, 1–33. Chicago: University of Chicago Press.

Jevons, William Stanley. 1866. *The Coal Question: An Inquiry Concerning the Progress of the Nation, and the Probable Exhaustion of our Coal-Mines.* London: Macmillan and Co.

Jha, Alok. 2013. "Astronomer Royal Calls for 'Plan B' to Prevent Runaway Climate Change." *The Guardian*, September 11. https://www.theguardian.com/science /2013/sep/11/astronomer-royal-global-warming-lord-rees.

Johnson, Scott K. 2015. "Shade the Planet? The Dangers Are in the Details." *Ars Technica*: February 18. http://arstechnica.com/science/2015/02/shade-the-planet -the-dangers-are-in-the-details/.

Jones, Caroline. 2016. "The New Cold War: The Political Problem of Geoengineering." *Brown Political Review*, May 2. https://brownpoliticalreview.org/2016/05/the-new -cold-war-the-political-problem-of-geoengineering/.

Jospe, Christophe. 2016. "The Good, Bad, and Ugly Approaches to Geoengineering." *Slate*: January 25. http://www.slate.com/articles/technology/future_tense/2016/01/ good_bad_and_ugly_approaches_to_geoengineering.html.

Kahan, Dan M. 2015. "You Can Change the Minds of Climate Change Skeptics. Here's How." *The Washington Post*: February 23. https://www.washingtonpost .com/blogs/monkey-cage/wp/2015/02/23/you-can-change-the-minds-of-climate -change-skeptics-heres-how/.

Keim, Brandon and Terry Macalister. 2015. "Shell's Arctic Oil Rig Departs Seattle as 'Kayaktivists' Warn of Disaster." *The Guardian*, June 15. https://www.theguardian .com/us-news/2015/jun/15/seattle-kayak-activists-detained-blocking-shell-arctic -oil-rig.

Keith, David. 2001. "Geoengineering." In *Encyclopedia of Global Change: Environmental Change and Human Society*, edited by Andrew Goudie, 495–502. Oxford: Oxford University Press.

Keith, David. 2013. *A Case for Climate Engineering*. Cambridge, MA: MIT Press.

Keith, David. 2016. "Why We Should Research Solar Geoengineering Now." *Slate*, January 19. https://slate.com/technology/2016/01/solar-geoengineering-is-not-a -quick-fix.html.

Keith, David. 2020. "Letter and Statement Regarding Financials and Conflict of Interest." Provided to SCoPEx Advisory Committee.

Keith Group. 2021. "Fund for Innovative Climate and Energy Research." Accessed February 23, 2021. https://keith.seas.harvard.edu/FICER.

Keutsch, Frank. 2020. "Advisory Committee Updates." Sent to SCoPEx email list August 3, 2020.

Keutsch Group at Harvard. 2020. "SCoPEx." Accessed October 27, 2020. https:// projects.iq.harvard.edu/keutschgroup/scopex.

Keutsch Group at Harvard. 2021. "SCoPEx: Stratospheric Controlled Perturbation Experiment." Accessed November 8, 2021. https://www.keutschgroup.com/ scopex.

Kintisch, Eli. 2010. *Hack the Planet: Science's Best Hope—or Worst Nightmare—for Averting Climate Catastrophe*. Hoboken: Wiley.

Kirby, Alex. 2018. "Solar Geoengineering 'Too Uncertain to Go Ahead Yet,' Expert Cautions." *Climate News Network*, April 6.

Kirsch, Scott and Don Mitchell. 1998. "Earth-Moving as the 'Measure of Man': Edward Teller, Geographical Engineering, and the Matter of Progress." *Social Text* 54 (Spring): 100–134.

Klein, Naomi. 2014. *This Changes Everything: Capitalism vs. The Climate*. New York: Simon and Schuster.

Krieger, Lisa M. 2015. "San Jose: Scientists Call for 'Geoengineering' Tests to Find Ways to Cool the Planet." *San Jose Mercury*, February 14. https://www.mercu- rynews.com/2015/02/14/san-jose-scientists-call-for-geoengineering-tests-to-find -ways-to-cool-the-planet/.

Krohn, Wolfgang and Peter Weingart. 1987. "Commentary: Nuclear Power as a Social Experiment-European Political 'Fall Out' from the Chernobyl Meltdown." *Science, Technology, & Human Values* 12 (2): 52–58.

Lakoff, George and Mark Johnson. 1980. *Metaphors We Live By*. Chicago: University of Chicago Press.

Laurance, William F., Susan G. Laurance and Patricia Delamonica. 1998. "Tropical Forest Fragmentation and Greenhouse Gas Emissions." *Forest Ecology and Management* 110 (1): 173–180.

Laurance, William F., Heraldo L. Vasconcelos and Thomas E. Lovejoy. 2000. "Forest Loss and Fragmentation in the Amazon: Implications for Wildlife Conservation." *Oryx* 34 (1): 39–45.

Lavelle, Marianne and David Hasemyer. 2017. "Instrument of Power: How Fossil Fuel Donors Shaped the Anti-Climate Agenda of a Powerful Congressional Committee." *InsideClimate News*, December 5. https://insideclimatenews.org/news/05122017/lamar-smith-congress-climate-change-fossil-fuel-industry-house-science-committee/.

Lenton, Timothy M. 2011. "Early Warning of Climate Tipping Points." *Nature Climate Change* 1: 201–209.

Levidow, Les and Susan Carr. 2007. "GM Crops on Trial: Technological Development as a Real-World Experiment." *Futures* 39 (4): 408–431.

Levitan, Dave. 2013. "Quick-Change Planet: Do Global Climate Tipping Points Exist?" *Scientific American*, March 25. https://www.scientificamerican.com/article/do-global-tipping-points-exist/.

Lindseth, Brian S. 2013. "From Radioactive Fallout to Environmental Critique: Ecology and the Politics of Cold War Science." Ph.D. Dissertation, University of California, San Diego.

Liptak, Adam and Coral Davenport. 2016. "Supreme Court Deals Blow to Obama's Efforts to Regulate Coal Emissions." *The New York Times*, February 9, 2016. https://www.nytimes.com/2016/02/10/us/politics/supreme-court-blocks-obama-epa-coal-emissions-regulations.html.

Lohr, Steve. 2009. "Pressing the Case for Geoengineering." *The New York Times*, June 22. https://green.blogs.nytimes.com/2009/06/22/pressing-the-case-for-geoengineering/

Long, Jane C. S. and Dane Scott. 2013. "Vested Interests and Geoengineering Research." *Issues in Science and Technology* Spring 2013: 45–52.

Lukacs, Martin. 2012. "US Geoengineers to Spray Sun-Reflecting Chemicals from Balloon." *The Guardian*, July 17. http://www.theguardian.com/environment/2012/jul/17/us-geoengineers-spray-sun-balloon.

Lukacs, Martin. 2017. "Trump Presidency 'Opens Door' to Planet Hacking Geoengineer Experiments." *The Guardian*, March 27. https://www.theguardian.com/environment/true-north/2017/mar/27/trump-presidency-opens-door-to-planet-hacking-geoengineer-experiments.

Lukacs, Martin, Suzanne Goldenberg and Adam Vaughan. 2013. "Russia Urges UN Climate Report to Include Geoengineering." *The Guardian*, September 19. http://www.theguardian.com/environment/2013/sep/19/russia-un-climate-report-geoengineering.

Lukes, Steven. 2005. *Power: A Radical View*. New York: Palgrave Macmillan.

Luokkanen, Matti, Suvi Huttunen and Mikael Hildén. 2014. "Geoengineering, News Media and Metaphors: Framing the Controversial." *Public Understanding of Science* 23 (8): 966–981.

Macnaghten, Phil and Bronislaw Szerszynski. 2013. "Living the Global Social Experiment: An Analysis of Public Discourse on Solar Radiation Management and its Implications for Governance." *Global Environmental Change* 23 (2): 465–474.

Malm, Andreas. 2018. *The Progress of this Storm: Nature and Society in a Warming World.* London: Verso.

Malm, Andreas. 2021. "Planning the Planet: Geoengineering Our Way Out of and Back into a Planned Economy." In *Has It Come to This? The Promises and Perils of Geoengineering on the Brink*, edited by J. P. Saspinski, Holly Jean Buck and Andreas Malm, 143–162. New Brunswick: Rutgers University Press.

Marchetti, Cesare. 1977. "On Geoengineering and the CO_2 Problem." *Climatic Change* 1: 59–68.

Markusson, Nils. 2013. "Tensions in Framings of Geoengineering: Constitutive Diversity and Ambivalence." *Climate Geoengineering Governance Working Paper Series* 003.

Mathesius, Sabine, Matthias Hofmann, Ken Caldeira and Hans Joachim Schellnhuber. 2015. "Long-Term Response of Oceans to CO_2 Removal from the Atmosphere." *Nature Climate Change* 5 (12): 1107–1113.

Mathiesen, Karl. 2015. "Is Geoengineering a Bad Idea?" *The Guardian*, February 11. https://www.theguardian.com/environment/2015/feb/11/is-geoengineering-a-bad-idea-climate-change.

Matthews, H. Damon and Ken Caldeira. 2007. "Transient Climate-Carbon Simulations of Planetary Geoengineering." *Proceedings of the National Academy of Sciences* 104: 9949–9954.

Maynard, Andrew. 2009. "Geoengineering Options: Balancing Effectiveness and Safety." Accessed March 15, 2017, 2017. http://2020science.org/2009/09/01/geoengineering-options-balancing-effectiveness-and-safety/.

McCright, Aaron M. and Riley E. Dunlap. 2010. "Anti-reflexivity: The American Conservative Movement's Success in Undermining Climate Science and Policy." *Theory, Culture & Society* 27 (2–3): 100–133.

McNerney, Jerry. 2017a. "Rep. McNerney Introduces Groundbreaking Geoengineering Bill." Accessed April 9, 2018. https://mcnerney.house.gov/media-center/press-releases/rep-mcnerney-introduces-groundbreaking-geoengineering-bill.

McNerney, Jerry. 2017b. "H.R. 4586 - Geoengineering Research Evaluation Act of 2017." *House of Representatives.* https://www.congress.gov/bill/115th-congress/house-bill/4586.

McNerney, Jerry. 2019. "H.R. 5519 - Atmospheric Climate Intervention Research Act." *House of Representatives.* https://www.congress.gov/bill/116th-congress/house-bill/5519/all-info.

Mehan, Hugh. 1987. "Language and Power in Organizational Process." *Discourse Processes* 10 (4): 291–301.

Mehan, Hugh. 2000. "Beneath the Skin and Between the Ears: A Case Study in the Politics of Representation." In *Schooling the Symbolic Animal: Social and Cultural Dimensions of Education*, edited by Bradley A. U. Levinson, 259–279. Lanham: Rowman & Litttlefield.

Mehan, Hugh, Charles E. Nathanson and James M. Skelly. 1990. "Nuclear Discourse in the 1980s: The Unravelling Conventions of the Cold War." *Discourse & Society* 1 (2): 133–165.

Mehan, Hugh and John Wills. 1988. "A Nurturing Voice in the Nuclear Arms Debate." *Social Problems* 35 (4): 363–383.

Mehling, Michael A. and Antto Vihma. 2017. "'Mourning for America': Donald Trump's Climate Change Policy." In *FIIA Analysis*. Helsinki: The Finnish Institute of International Affairs. https://papers.ssrn.com/sol3/papers.cfm?abstract _id=3051901.

Melillo, Jerry M., Terese (T.C.) Richmond and Gary W. Yohe, eds. 2014. *Climate Change Impacts in the United States: The Third National Climate Assessment*. U.S. Global Change Research Program.

Mervis, Jeffrey. 2014. "At House Science Panel Hearing, Sarcasm Rules." *Science*, March 28. https://www.science.org/content/article/house-science-panel-hearing -sarcasm-rules.

Mol, Arthur P. J. 2000. "The Environmental Movement in an Era of Ecological Modernisation." *Geoforum* 31: 45–56.

Mol, Arthur P. J. and David A. Sonnenfeld. 2000. "Ecological Modernisation Around the World: An Introduction." *Environmental Politics* 9 (1): 1–14.

Mol, Arthur P. J. and Gert Spaargaren. 2000. "Ecological Modernisation Theory in Debate: A Review." *Environmental Politics* 9 (1): 17–49.

Möller, Ina. 2021. "Winning Hearts and Minds? Explaining the Rise of the Geoengineering Idea." In *Has It Come to This? The Promises and Perils of Geoengineering on the Brink*, edited by J. P. Saspinski, Holly Jean Buck and Andreas Malm, 21–33. New Brunswick: Rutgers University Press.

Morrow, David R. 2014. "Ethical Aspects of the Mitigation Obstruction Argument Against Climate Engineering Research." *Philosophical Transactions of the Royal Society of London A: Mathematical, Physical and Engineering Sciences* 372 (2031): 20140062.

Morton, Oliver. 2015. "Will Our Grandchildren Say That We Changed the Earth Too Little?" Interview by Ross Andersen: *The Atlantic*. https://www.theatlantic .com/science/archive/2015/11/its-time-to-start-talking-about-geoengineering /414283/.

Mukerji, Chandra. 1990. *A Fragile Power: Scientists and the State*. Princeton: Princeton University Press.

NASEM, National Academies of Sciences, Engineering, and Medicine. 2019. *Negative Emissions Technologies and Reliable Sequestration: A Research Agenda*. Washington, DC: The National Academies Press.

NASEM, National Academies of Sciences, Engineering, and Medicine. 2021. *Reflecting Sunlight: Recommendations for Solar Geoengineering Research and Research Governance*. Washington, DC: The National Academies Press.

National Academies Webinar. 2021. *Reflecting Sunlight: Recommendations for Solar Geo-engineering Research and Research Governance- Report Release Webinar*. The National Academies Press. Retrieved February 20, 2022, https://vimeo.com /530878274.

National Research Council. 2015a. *Climate Intervention: Carbon Dioxide Removal and Reliable Sequestration*. Pre-publication Copy, Washington, DC: National Academies Press.

National Research Council. 2015b. *Climate Intervention: Reflecting Sunlight to Cool Earth*. Pre-publication Copy, Washington, DC: National Academies Press.

National Research Council, Committee on the Status of and Future Directions in U.S. Weather Modification Research and Operations, Board on Atmospheric Sciences and Climate, Division on Earth and Life Studies and National Research Council of the National Academies. 2003. *Critical Issues in Weather Modification Research*. Washington, DC: National Academies Press.

Nerlich, Brigitte and Rusi Jaspal. 2012. "Metaphors We Die by? Geoengineering, Metaphors, and the Argument from Catastrophe." *Metaphor and Symbol* 27 (2): 131–147.

Nocera, Joe. 2015. "Chemo for the Planet." *The New York Times*, May 19. https://www.nytimes.com/2015/05/19/opinion/joe-nocera-chemo-for-the-planet.html.

Norgaard, Kari Marie. 2006. "'People Want to Protect Themselves a Little Bit': Emotions, Denial, and Social Movement Nonparticipation." *Sociological Inquiry* 76 (3): 372–396.

North, Douglass Cecil. 1990. *Institutions, Institutional Change, and Economic Performance*. Cambridge, New York: Cambridge University Press.

Nowotny, Helga, Peter B. Scott and Michael T. Gibbons. 2001. *Re-Thinking Science: Knowledge and the Public in an Age of Uncertainty*. Cambridge: Polity.

O'Gorman, Ned and Kevin Hamilton. 2011. "At the Interface: The Loaded Rhetorical Gestures of Nuclear Legitimacy and Illegitimacy." *Communication and Critical/Cultural Studies* 8 (1): 41–66.

Office of the Clerk. 2018. *Committee FAQs*. U.S. House of Representatives. Accessed January 27, 2018. http://clerk.house.gov/committee_info/commfaq.aspx.

Oreskes, Naomi. 2004. "The Scientific Consensus on Climate Change." *Science* 306: 1686.

Oreskes, Naomi and Erik M. Conway. 2010. *Merchants of Doubt: How a Handful of Scientists Obscured the Truth on Issues from Tobacco Smoke to Global Warming*. New York: Bloomsbury Press.

Owen, Richard. 2014. "Solar Radiation Management and the Governance of Hubris." In *Geoengineering of the Climate System, Vol. 38, Issues in Environmental Science and Technology*, edited by R. E. Hester and R. M. Harrison, 212–248. London: The Royal Society of Chemistry.

Pappas, Stephanie. 2013. "Incredible Technology: How to Engineer the Climate." *LiveScience*: July 1. http://www.livescience.com/37865-how-to-engineer-the-climate.html.

Parkinson, Claire L. 2010. *Coming Climate Crisis?: Consider the Past, Beware the Big Fix*. Lanham: Rowman & Littlefield.

Parry, Sarah. 2009. "Stem Cell Scientists' Discursive Strategies for Cognitive Authority." *Science as Culture* 18 (1): 89–114.

Perrow, Charles. 2011. *The Next Catastrophe: Reducing Our Vulnerabilities to Natural, Industrial, and Terrorist Disasters.* Princeton: Princeton University Press.

Petersen, Arthur. 2014. "The Emergence of the Geoengineering Debate within the IPCC." *Case Study, Geoengineering Our Climate Working Paper and Opinion Article Series.*

Pimm, S. L., C. N. Jenkins, R. Abell, T. M. Brooks, J. L. Gittleman, L. N. Joppa, P. H. Raven, C. M. Roberts and J. O. Sexton. 2014. "The Biodiversity of Species and their Rates of Extinction, Distribution, and Protection." *Science* 344 (6187): 1246752.

Plumer, Brad. 2014. "One Problem with Geoengineering: Once You Start, You Can't Really Stop." *The Washington Post*, January 2. https://www.washingtonpost.com/news/wonk/wp/2014/01/02/one-problem-with-geoengineering-once-you-start-you-cant-ever-stop/.

Plumer, Brad and Christopher Flavelle. 2021. "Businesses Aim to Pull Greenhouse Gases From the Air. It's a Gamble." *The New York Times*, October 10. https://www.nytimes.com/2021/01/18/climate/carbon-removal-technology.html.

Plumer, Brad and Henry Fountain. 2020. "Trump Administration Finalizes Plan to Open Arctic Refuge to Drilling." *New York Times*: August 17. www.nytimes.com/2020/08/17/climate/alaska-oil-drilling-anwr.html.

Polanyi, Karl. [1944] 2001. *The Great Transformation: The Political and Economic Origins of Our Time.* Boston: Beacon Press.

Poortinga, Wouter and Nick Pidgeon. 2007. "Public Perceptions of Agricultural Biotechnology in the UK: The Case of Genetically Modified Food." In *The Media, the Public and Agricultural Biotechnology*, edited by Dominique Brossard, James Shanahan and T. Clint Nesbitt, 21–56. Wallingford: CAB International.

Popovich, Nadja, Livia Albeck-Ripka and Kendra Pierre-Louis. 2020. "The Trump Administration Is Reversing 100 Environmental Rules. Here's the Full List." *The New York Times*, July 15 (updated January 20, 2021). https://www.nytimes.com/interactive/2020/climate/trump-environment-rollbacks-list.html.

Porter, Eduardo. 2016. "Next Supreme Court Justice Will be Crucial to Climate Change." *The New York Times*, February 16. https://www.nytimes.com/2016/02/17/business/economy/next-supreme-court-justice-will-be-crucial-to-climate-change.html.

Rasch, Philip J, Simone Tilmes, Richard P Turco, Alan Robock, Luke Oman, Chih-Chieh Chen, Georgiy L Stenchikov and Rolando R Garcia. 2008. "An Overview of Geoengineering of Climate Using Stratospheric Sulphate Aerosols." *Philosophical Transactions of the Royal Society of London A: Mathematical, Physical and Engineering Sciences* 366 (1882): 4007–4037.

Rayner, Steve, Clare Heyward, Tim Kruger, Nick Pidgeon, Catherine Redgwell and Julian Savulescu. 2013. "The Oxford Principles." *Climatic Change* 121: 499–512.

Readfearn, Graham. 2014. "Geoengineering the Earth's Climate Sends Policy Debate Down a Curious Rabbit Hole." *The Guardian*, August 3. https://www.theguardian.com/environment/planet-oz/2014/aug/04/geoengineering-the-earths-climate-sends-policy-debate-down-a-curious-rabbit-hole.

Rees, Martin. 2009. "Forward." In *Geoengineering the Climate: Science, Governance and Uncertainty.* London: The Royal Society.

Rees, Martin. 2013. "Transcript of Speech Given at the British Science Festival in Newcastle on September 12." *The Conversation.* Accessed September 26, 2013. theconversation.com/astronomer-royal-on-science-environment-and-the-future-18162.

Regan, Sheila. 2021. "'It's Cultural Genocide': Inside the Fight to Stop a Pipeline on Tribal Lands." *The Guardian*, February 19. https://www.theguardian.com/us-news /2021/feb/19/line-3-pipeline-ojibwe-tribal-lands.

Revkin, Andrew C. 2009. "Among Climate Scientists, a Dispute Over 'Tipping Points.'" *The New York Times*, March 28. https://www.nytimes.com/2009/03/29/ weekinreview/29revkin.html.

Revkin, Andrew C. 2015. "Why Hacking the Atmosphere Won't Happen Any Time Soon." *The New York Times*, February 12. https://dotearth.blogs.nytimes.com/2015 /02/12/why-hacking-the-atmosphere-wont-happen-any-time-soon/.

Revkin, Andrew C. 2016. "Can Humans Go From Unintended Global Warming to Climate By Design?" *The New York Times*, October 18. https://dotearth.blogs .nytimes.com/2016/10/18/can-humans-go-from-unintended-global-warming-to -climate-by-design/.

Robock, Alan. 2000. "Volcanic Eruptions and Climate." *Reviews of Geophysics* 38 (2): 191–219.

Robock, Alan. 2008. "20 Reasons Why Geoengineering May Be a Bad Idea." *Bulletin of the Atomic Scientists* 64 (2): 14–18, 59.

Robock, Alan, Martin Bunzl, Ben Kravitz and Georgiy L. Stenchikov. 2010. "A Test for Geoengineering?" *Science* 327 (5965): 530–531.

Royal Society. 2009. *Geoengineering the Climate: Science, Governance and Uncertainty.* London: The Royal Society.

Royal Society. 2016. "History." The Royal Society. Accessed April 21, 2016, https:// royalsociety.org/about-us/history/.

Royal Society and Royal Academy of Engineering. 2018. *Greenhouse Gas Removal.* London: Royal Society.

Rubin, Beatrix P. 2008. "Therapeutic Promise in the Discourse of Human Embryonic Stem Cell Research." *Science as Culture* 17 (1): 13–27.

Samenow, Jason, Kasha Patel, Hannah Dormido and Laris Karklis. 2021. "How Tropical Storms and Hurricanes Have Hit U.S. Shores with Unparalleled Frequency." *The Washington Post*, September 29. https://www.washingtonpost .com/weather/2021/09/29/record-us-hurricane-landfalls-climate/.

Sample, Ian. 2015. "Spy Agencies Fund Climate Research in Hunt for Weather Weapon, Scientist Fears." *The Guardian*, February 15. https://www.theguardian.com/environ- ment/2015/feb/15/spy-agencies-fund-climate-research-weather-weapon-claim.

Saspinski, J. P., Holly Jean Buck and Andreas Malm, eds. 2021. *Has It Come to This? The Promises and Perils of Geoengineering on the Brink.* New Brunswick: Rutgers University Press.

Schnaiberg, Allan. 1980. *The Environment: From Surplus to Scarcity.* New York: Oxford University Press.

Schneider, Linda and Lili Fuhr. 2021. "Defending a Failed Status Quo: The Case against Geoengineering from a Civil Society Perspective." In *Has It Come to This? The Promises and Perils of Geoengineering on the Brink*, edited by J. P. Saspinski, Holly Jean Buck and Andreas Malm, 50–68. New Brunswick: Rutgers University Press.

SCoPEx Advisory Committee. 2020a. "Financial Review." Accessed February 21, 2021. https://scopexac.com/financial-review/.

SCoPEx Advisory Committee. 2020b. "Financial Review Update from the Advisory Committee." Email sent to SCoPEx email list November 25, 2020.

Selin, Cynthia. 2007. "Expectations and the Emergence of Nanotechnology." *Science, Technology, & Human Values* 32 (2): 196–220.

Sikka, Tina. 2012. "A Critical Discourse Analysis of Geoengineering Advocacy." *Critical Discourse Studies* 9 (2): 163–175.

Solar Radiation Management Governance Initiative. 2011. "Solar Radiation Management: The Governance of Research." Accessed November 18, 2013. http://www.srmgi.org/files/2012/01/DES2391_SRMGI-report_web_11112.pdf.

Soltani, Alireza, Benedetto De Martino and Colin Camerer. 2012. "A Range-Normalization Model of Context-Dependent Choice: A New Model and Evidence." *PLoS Computational Biology* 8 (7): e1002607.

Specter, Michael. 2012. "The Climate Fixers." *The New Yorker*, May 14, 96–103.

Stilgoe, Jack. 2015. *Experiment Earth: Responsible Innovation in Geoengineering*. Abingdon, Oxfordshire: Routledge.

Stilgoe, Jack. 2016. "Geoengineering as Collective Experimentation." *Science and Engineering Ethics* 22: 851–869.

Stirling, Andrew. 2008. "'Opening Up' and 'Closing Down': Power, Participation, and Pluralism in the Social Appraisal of Technology." *Science, Technology, & Human Values* 33 (2): 262–294.

Stirling, Andrew. 2014. *Emancipating Transformations: From Controlling 'the transition' to Culturing Plural Radical Progress*. STEPS Working Paper 64, Brighton: STEPS Centre.

Stuart, Diana, Ryan Gunderson and Brian Petersen. 2020. "Carbon Geoengineering and the Metabolic Rift: Solution or Social Reproduction?" *Critical Sociology* 46 (7–8): 1233–1249.

Sukhodolov, Timofei, Jian-Xiong Sheng, Aryeh Feinberg, Bei-Ping Luo, Thomas Peter, Laura Revell, Andrea Stenke, Debra K. Weisenstein and Eugene Rozanov. 2018. "Stratospheric Aerosol Evolution after Pinatubo Simulated with a Coupled Size-Resolved Aerosol–Chemistry–Climate Model, SOCOL-AERv1.0." *Geoscientific Model Development* 11 (7): 2633–2647.

Sumner, Thomas. 2015. "Geoengineering Is World's Last Hope, New Book Argues." *Science News*, November 15. https://www.sciencenews.org/article/geoengineering-worlds-last-hope-new-book-argues.

Szerszynski, Bronislaw, Matthew Kearnes, Phil Macnaghten, Richard Owen and Jack Stilgoe. 2013. "Why Solar Radiation Management Geoengineering and Democracy Won't Mix." *Environment and Planning A* 45: 2809–2816.

The Center for Responsive Politics. 2018. "Rep. Lamar Smith - Texas District 21." Accessed February 3, 2018. https://www.opensecrets.org/members-of-congress/industries?cycle=Career&cid=N00001811.

The Economist Staff. 2010a. "Geoengineering: Lift-off." *The Economist*, November 4.

The Economist Staff. 2010b. "We All Want to Change the World." *The Economist*, March 31. http://www.economist.com/node/15814427.

The Economist Staff. 2013. "Stopping a Scorcher: The Controversy Over Manipulating Climate Change." *The Economist*, November 23. http://www.economist.com/news/books-and-arts/21590347-controversy-over-manipulating-climate-change-stopping-scorcher.

Thorpe, Charles. 2013. "Artificial Life on a Dead Planet." In *Media Studies Futures, Vol. VI, The International Encyclopedia of Media Studies*, edited by Kelly Gates, 615–647. Oxford: Blackwell Publishing.

Thorpe, Charles. 2016. *Necroculture*. New York: Palgrave Macmillan.

Thorpe, Charles and Brynna Jacobson. 2013. "Life Politics, Nature and the State: Giddens' Sociological Theory and *The Politics of Climate Change*." *The British Journal of Sociology* 64 (1): 99–122.

Thorpe, Charles and Brynna Jacobson. 2021. "Abstract Life, Abastract Labor, Abstract Mind." *Research in Political Economy* 35: 59–105.

Toulmin, Stephen E. 1958. *The Uses of Argument*. Cambridge: Cambridge University Press.

Tribes, Klamath. 2017. "Klamath Tribes News and Events: The Official Klamath Tribes News Page." Accessed February 20, 2021. https://klamathtribes.org/news/you-can-help-us-stop-the-pipeline-in-oregon/.

Twomey, Sean. 1977. "The Influence of Pollution on the Shortwave Albedo of Clouds." *Journal of Atmospheric Science* 34: 1149–1152.

U.S. Government Publishing Office. 1999. "About Congressional Hearings." Accessed January 27, 2018. https://www.gpo.gov/help/about_congressional_hearings.htm.

United States Climate Alliance. 2020. "About Us." Accessed July 20, 2020. http://www.usclimatealliance.org/about-us.

United States House of Representatives. 2017. "The House Explained." Accessed January 27, 2018. https://www.house.gov/the-house-explained.

Upton, John. 2015. "Geoengineering Is Too Risky, Scientists Warn Paris COP21 Negotiators." *Alternet*, December 8. https://www.alternet.org/2015/12/geoengineering-too-risky-scientists-warn-paris-cop21-negotiators/.

Urry, John. 2011. *Climate Change and Society*. Cambridge: Polity Press.

Venkataraman, Bina. 2016. "Experimenting with Geoengineering Could Have Unintended Consequences." *Slate*: January 12. http://www.slate.com/articles/technology/future_tense/2016/01/experimenting_with_geoengineering_could_have_unintended_consequences.html.

Victor, Daniel. 2021. "Flooding in China Kills 21, as Thousands Escape to Shelters." *The New York Times*, August 13. https://www.nytimes.com/2021/08/13/world/asia/china-flooding-evacuations.html.

Victor, David. 2019. "Governing the Deployment of Geoengineering: Institutions, Preparedness, and the Problem of Rogue Actors." In *Governance of the Deployment of Solar Geoengineering, Harvard Project on Climate Agreements*, edited by Robert N. Stavins and Robert C. Stowe. Cambridge: Harvard University Press.

Victor, David G. 2011. *Global Warming Gridlock: Creating More Effective Strategies for Protecting the Planet*. Cambridge: Cambridge University Press.

Victor, David G., M. Granger Morgan, Jay Apt, John Steinbruner and Katharine Ricke. 2009. "The Geoengineering Option: A Last Resort Against Global Warming?" *Foreign Affairs* 88 (2): 64–76.

Vidal, John. 2018. "How Bill Gates Aims to Clean up the Planet." *The Guardian*, February 4. https://www.theguardian.com/environment/2018/feb/04/carbon -emissions-negative-emissions-technologies-capture-storage-bill-gates.

Virgin. 2021. "Virgin Earth Challenge." Accessed February 23, 2021 https://www .virgin.com/about-virgin/virgin-group/news/virgin-earth-challenge.

Watts, Jonathan. 2021. "Climate Scientists Shocked by Scale of Floods in Germany." *The Guardian*, July 16. https://www.theguardian.com/environment/2021/jul/16/ climate-scientists-shocked-by-scale-of-floods-in-germany.

Watzlawick, Paul. 2011 [1984]. "Self-Fulfilling Prophecies." In *The Production of Reality: Essays and Readings on Social Interaction*, edited by Jodi O'Brien, 392–403. Los Angeles: Sage.

Weisenthal, Joe. 2009. "The 10 Most-Respected Global Warming Skeptics." *Business Insider*, July 30. https://www.businessinsider.com/the-ten-most-important-climate -change-skeptics-2009-7.

Welch, Craig. 2015. "There's a Good and a Bad Way to 'Geoengineer' the Planet." *National Geographic*, February 9. https://www.nationalgeographic.com /science/article/150210-national-academy-geoengineering-report-climate-change -environment.

Welsh, Ian. 2000. *Mobilising Modernity: The Nuclear Moment*. New York: Routledge.

White House. 2021a. *Executive Order on Protecting Public Health and the Environment and Restoring Science to Tackle the Climate Crisis*, January 20. https://www.whitehouse.gov/briefing-room/presidential-actions/2021/01/20/ executive-order-protecting-public-health-and-environment-and-restoring-science -to-tackle-climate-crisis/.

White House. 2021b. "Fact Sheet: President Biden Takes Executive Actions to Tackle the Climate Crisis at Home and Abroad, Create Jobs, and Restore Scientific Integrity Across Federal Government." January 27. https://www.whitehouse.gov /briefing-room/statements-releases/2021/01/27/fact-sheet-president-biden-takes -executive-actions-to-tackle-the-climate-crisis-at-home-and-abroad-create-jobs -and-restore-scientific-integrity-across-federal-government/.

Wigglesworth, Alex. 2021. "Climate Change is Now the Main Driver of Increasing Wildfire Weather, Study Finds." *Los Angeles Times*, November 1. https://www .latimes.com/california/story/2021-11-01/climate-change-is-now-main-driver-of -wildfire-weather.

Wilsdon, James, Brian Wynne and Jack Stilgoe. 2004. *The Public Value of Science: Or how to Ensure that Science Really Matters.* London: Demos.

Wuebbles, D. J., D. W. Fahey, K. A. Hibbard, D. J. Dokken, B. C. Stewart and T. K. Maycock (Eds.). 2017. "Climate Science Special Report: Fourth National Climate Assessment." Vol. I. Washington, DC: U.S. Global Change Research Program.

XPRIZE Foundation. 2021. "XPrize: Prizes." Accessed February 23, 2021. https://www.xprize.org/prizes.

Yang Campaign. 2019. "Policy: Geoengineering." Accessed February 27, 2021. https://2020.yang2020.com/policies/geoengineering/.

York, Richard. 2021. "Geoengineering and Imperialism." In *Has It Come to This? The Promises and Perils of Geoengineering on the Brink*, edited by J. P. Saspinski, Holly Jean Buck and Andreas Malm, 179–188. New Brunswick: Rutgers University Press.

York, Richard and Eugene A. Rosa. 2003. "Key Challenges to Ecological Modernization Theory." *Organization & Environment* 16 (3): 273–288.

Index

Page references for figures and tables are italicized.

About the Author

Brynna Jacobson teaches as part-time faculty in the Department of Sociology at the University of San Francisco. She received her doctorate degree in sociology from the University of California, San Diego. She also holds a master's degree from the University of San Diego and a bachelor's degree from the University of California, Berkeley. Her research interests include environmental sociology, political economy, law and society, and discourse analysis. She lives in San Francisco with her family.

www.ingramcontent.com/pod-product-compliance
Lightning Source LLC
Chambersburg PA
CBHW022304280326
41932CB00010B/980